A BEEKEEPER'S YEAR

A Beekeeper's Year

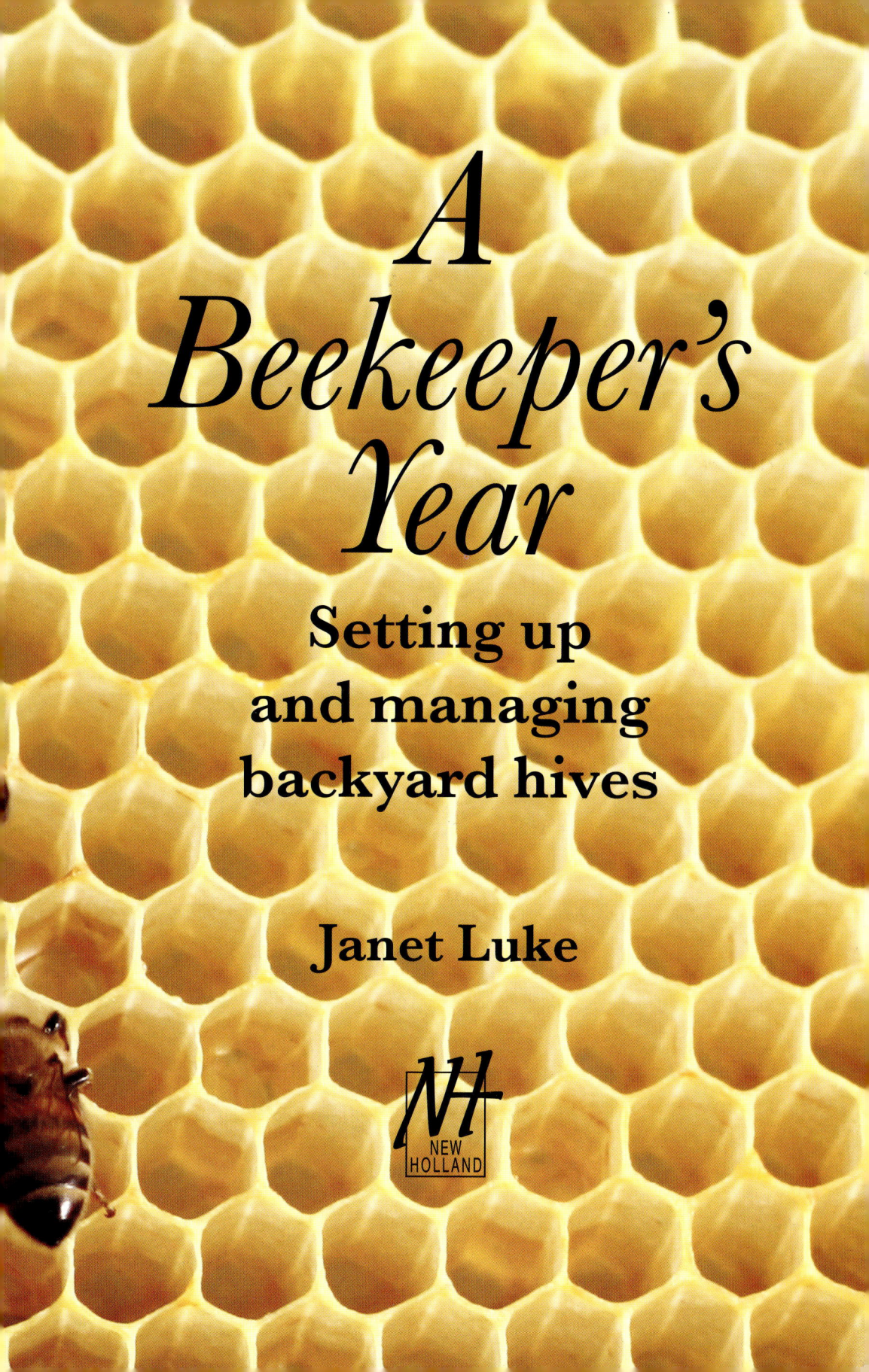

Setting up and managing backyard hives

Janet Luke

NEW HOLLAND

CONTENTS

Introduction 6

SECTION I

CHAPTER 1
WHY KEEP BEES? 11

Stings 13
Avoiding stings 15
Time commitment 16
A rough guide to time commitment
    for a backyard hive 17
What does it take to be a great
    beekeeper? 18
How hard is it to keep bees? ... 19
What about children and pets? ... 19

CHAPTER 2
GETTING STARTED 21

Do your research 22
Beekeeping clubs 22
Developing your own beekeeping ethos 22
Space requirements 24
Legal requirements 24
Getting bees 25

CHAPTER 3
THE LIFE OF BEES 29

The honey bee world 30
The mechanisms of a hive 37

CHAPTER 4
YOUR APIARY SITE 57

Moving a hive to a new position
    in your garden 58
What makes a good site for a hive? .. 59
Working the bees in your garden ... 62

CHAPTER 5
EQUIPMENT 65

Bee suit 66
Gloves 67
Hive tool 68
Smoker 69
Veil 73

SECTION II

CHAPTER 6
INTRODUCING OUR
THREE NEW BEEKEEPERS ... 77

Sarah and her Flow hive 78
What is a Flow hive? 79
Honey harvesting with the Flow hive ... 80
How the Flow hive started 80
Who is it made for? 81
Concerns 81
The pros and cons of the Flow hive ... 83
Eric and his Warré hive 84
The Warré hive 86
How the Warré hive started ... 87
The pros and cons of the Warré hive ... 88
Sonya and her Top Bar hive ... 90
The horizontal Top Bar hive ... 91
Who is it made for? 91
The pros and cons of the Top Bar hive ... 93

CHAPTER 7
GETTING BEES IN YOUR HIVE — 99

What is a swarm?	103
How to collect a swarm	106
Transferring a swarm into your hive	107
How to stop a swarm absconding from a new hive	108
To feed or not to feed a swarm	109
Monitoring your hive as they build comb	109
Varroa treatments for your new swarm	111
Building your own swarm traps	112
Where to place your bait hive	114
OMG! I've caught a swarm of bees in my bait hive—what do I do now?	114
Questions to ask the seller of a nucleus beehive:	116
To feed or not to feed?	117
Making a 2:1 sugar syrup feed	118
Transporting a nucleus colony	119
Transferring the nucleus colony into a hive	120
Obtaining a split	120
Placing a nucleus into a Flow or Langstroth brood box	124

CHAPTER 8
SEASONAL MANAGEMENT IN YOUR FIRST YEAR OF BEEKEEPING — 129

Spring management in your first year of beekeeping	130
How to inspect your new hive	130
For a Warré hive and a Flow hive	132
Conducting a Top Bar hive inspection	133
What are you looking at when you inspect a frame?	137
Important nectar and pollen sources for spring bees	142
Late spring–early summer: building up	143
What is a nectar flow and how to tell if one is on in your area?	143
Signs that a nectar flow is happening in your area	144
Can good-quality queens be raised by this emergency method?	150
Late summer–early autumn	154
Robbing	154
If you witness a robbing session, what can you do?	158
Extracting honey	159
How to harvest honey from a Flow hive	165
Why is it important to only harvest honey that has been capped?	169
Granulated honey	170
Early autumn management of your hive	170
Tutin toxic honey	170
Late autumn management of your hive	172
Preparing your hives for winter	175

CHAPTER 9
PEST MANAGEMENT — 179

Varroa mites	180
How to monitor for varroa levels within your hive	183
Using integrated pest management to control mites	185
Treatment-free beekeeping	187
Organic treatments for your hive	188
Organic treatments	190
Some of the organic treatment options	196
Pests and diseases of the honey bee hive	201
Diseases of the hive	204
Conclusion	212
Author Tips	213
Bibliography	215

INTRODUCTION

Keeping bees seems like a simple thing to do, after all haven't they been around for millions of years and are basically a wild insect? It used to be simple but over the years we have managed to complicate it in many ways and now it can be a real challenge to keep a hive of bees alive.

There are many books on bees and beekeeping and it can be very confusing to know where to start. Hopefully I can share some insight into alternative, more bee-focused ways to keep one or two hives in your garden. The focus of this book is to provide guidance for the hobby beekeeper. It does not follow conventional commercial beekeeping practices as I strongly feel that this type of beekeeping is not well suited to keeping healthy bees nor backyard beekeeping.

I find keeping bees fascinating and I am forever learning from the bees. If you are only interested in keeping bees to produce the maximum amount of honey then this book is probably not for you. My paradigm for keeping bees is about providing a home for my bees which is bee-focused rather than honey production-focused, in a way which is low-tech, sustainable and ethical. I enjoy keeping bees to aid pollination of my edible garden and to enjoy the very best local, raw, organic honey.

When I open my hive I often see something new. I would never trust a bee-keeper who states they know everything there is to know about beekeeping. In my mind this means that they have stagnated and stopped learning, are set in their ways and are not placed to learn new techniques or theories nor change their management. That statement "we just do it this way because we have always done it like this" is not good enough in this changing and volatile climate and environment we live in.

The modern form of beekeeping, in which 95 percent of bees are kept, I believe is aiding in the decline of bees. Bees are being worked too hard. Commercial hive designs force them to continuously collect and store honey, which is then taken by the beekeeper and replaced with white sugar. The bees' home is laced with toxic chemicals to treat diseases that the beekeeper spreads. Bees are moved and placed in monocrops and forced to pollinate plant species that provide very little nutrition for them. Through the use of pressed plastic or wax foundation, the colony is forced to live in harems with no male bees permitted, as the beekeeper's mentality is that a male bee does not collect honey so therefore it is useless.

As backyard beekeepers, we have the unique opportunity to shun the management system of a commercial operation. We can keep bees solely for the pleasure of seeing them flying from flower to flower in our garden and enjoying their soft hum and the sweet smell of ripening nectar as we pass the hive. All our management decisions can be about what is best for our bees rather than what is best for maximum honey production.

This book will take us on a journey following three new beekeepers who have chosen three different hive designs. Hopefully, you will gain helpful insights and information to set you on your own fascinating journey of becoming a backyard beekeeper.

Beekeeping had the dubious honour of becoming the first part of our system of industrial agriculture to actually fall apart. We blame the mites, the weather, diseases, the markets, consumers, the Chinese, other beekeepers, the Germans, the packers, the price of gas, anything to actually have to face up to what is really happening. We are losing the ability to take care of living things. Why? The old beekeeping is dying and the new one is struggling to be born. Are you going to the funeral or are you assisting with the birth, you have to choose.

– Kirk Webster

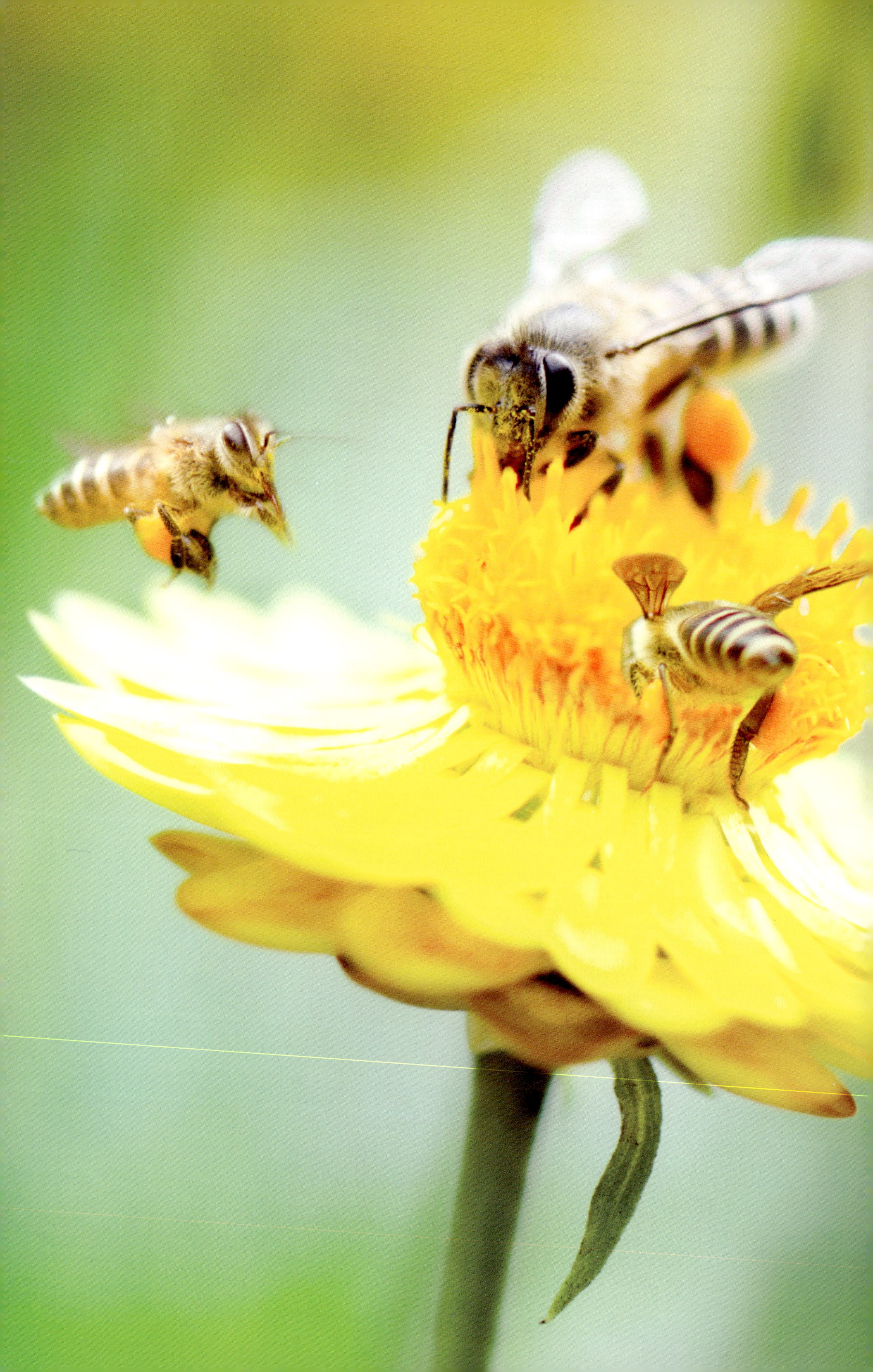

SECTION I

*Respect the colony as an organism
rather than a mechanism with interchangeable parts.*

– GUNTHER HAUK, *Toward Saving the Honeybee*

CHAPTER 1

WHY KEEP BEES?

Bees are an incredible insect. They evolved with flowers when dinosaurs walked the Earth. They have an intricate relationship with flowering plants and thus we, as the human race, depend on them and their skills in pollination to provide the food our survival hinges on. Along with the more obvious foods such as berries, fruit and vegetables, bees are also vital in industries such as dairy. Next time you are enjoying that aged Gouda remember that the bees helped make it. Without bees to pollinate the pasture, such as clover, for the dairy cows to feed off we could not enjoy the plethora of dairy products we do today.

Keeping bees brings us closer to nature. When you become a beekeeper you invariably become a weather person, environmentalist and a horticulturist by default. Beekeeping and the local environment are intricately linked.

There is a huge surge in interest in backyard beekeeping and this may be due to the related interest in backyard sustainability. It makes sense that when you start growing vegetables and planting fruit trees you quickly realize that bees are required for all your plans to come to fruition. Without the bee you have no pollination and without pollination you have little fruit or vegetables.

People are now aware of the trouble bees are in around the world and they want to play a part, however small, in trying to reverse the decline in numbers. By keeping one or two hives in your backyard garden you can help stem this decline.

Bees work tirelessly and we rely on them to create the food we enjoy eating. This bee is busy pollinating almond blossoms so we can enjoy almond nuts.

A honey bee busy pollinating an apple blossom.

Stings

Perhaps the biggest factor that puts people off keeping bees is the rear end of the bee. Let's face it, who in their right mind would choose to keep 50,000 stinging insects in a box in their garden? The cold hard truth of beekeeping is, yes, you will get stung. If you keep a hive in your garden you will have more bees in your space so there is an increased risk that other people may get stung by accident.

As a rule, honey bees are not aggressive. If they are flying from flower to flower in your garden they are not interested in you, only in gathering nectar and pollen for the hive. If a bee lands on you she is probably just needing a rest or the temperature has dropped and she is unable to fly. The sting is the bee's only defense mechanism and they prefer to only use it as a last resort. When a bee stings she will die, so she will never sting without a reason. When a bee stings a mammal, such as us or a dog, the stinger is torn out of her body as the barbs of the sting are stuck in our skin. The venom sac is pulled out when she tries to fly away. This disemboweling kills her.

It is the worker bee who stings. The drone (male bee) does not possess a stinger. The queen bee has a stinger but hers does not have barbs so can be used to sting several times. In general, the queen does not leave the hive, so her stinger is designed to fight other queen bees, not us mere mortals.

The primary reason a bee will sting is to protect the colony. If anything comes near and looks like it is about to rob them of their precious food stores or babies, it is fair game. A bee will also defend itself if something tries to crush it, such as someone stepping or sitting on the bee.

Bees do become more defensive as the colony builds in strength as there are more guard bees and more resources to protect. During summer, when there is lots of nectar to collect, the bees are too busy to become defensive. A small nucleus colony, which is busy caring for and raising a new queen, will generally not be defensive.

I have found that in early spring, colonies that have overwintered can be more aggressive than usual. This may be because these bees are all older and wiser, as there has not been a lot of brood rearing over winter.

Bees can be more aggressive if they are constantly harassed by outside forces. If you live in a country with raccoons or bears and they try to rob the hive, these bees can be defensive when the hive is opened. If loud fireworks are set off near the hive, or children have been throwing rocks on to the hive roof, this can also unsettle the hive.

If the nectar in your area dries up (this is known as a nectar dearth) the hive can become more defensive as they are under stress to find food for their developing young, and also to defend their hive and its resources from other robbing bees from neighboring colonies. Throw yourself into this mix and you are greeted by angry bees!

When you get stung, remove the stinger as quickly as possible as the venom sac will continue to pump venom into your circulation for up to 1 minute. Rather than using your fingers as a pincher to grasp the sting, scrape the sting out with your fingernail or hive tool. This action prevents you from squeezing the sac and pushing more venom into your body. Incidentally, a honey bee is the only sting to remain in the skin—hornet, bumblebee, and wasp stings do not remain in your skin.

When a bee stings you she releases an alarm pheromone, which may draw other bees to the sting site. Puffing that area with smoke will mask this chemical odor. The alarm pheromone smells like bananas, so it is a good idea not to eat bananas before any sort of beekeeping.

A sting will hurt. I always think it feels like a red-hot prickle. Many people think they are allergic to bees, but what they are experiencing when stung is a normal reaction. It is normal to experience pain, redness, swelling and itchiness. Apply white vinegar, ice or comfrey leaves to the sting site. To help with the annoying swelling and itchiness, use an over-the-counter antihistamine cream or tablet. I use a homeopathic remedy called Apis Mellifica, which comes as a small sugar tablet to dissolve under the tongue.

A life-threatening allergic reaction involves swelling around the neck and inside the throat, tongue swelling, dizziness, red swollen welts (hives) all over the body and shortness of breath. Any of these signs mean it is a medical emergency. Get medical help immediately by calling an ambulance and administering adrenaline, if you have it. This type of allergy occurs in about 1 percent of the population.

Beekeepers who encounter stings on a regular basis normally develop a lower sensitivity to bee venom. I personally have found that after beekeeping for over nine years I now do not swell at all and only have the bothersome itching for a few hours. Before I started keeping bees, if I got stung on my foot, my foot would swell to the size of a football and stay that way for a few days. Allergies are a strange beast. They can go one way or the other. You could be part of the unlucky minority and develop a severe allergy. In that case the safest option is to give up beekeeping. My two sons reacted badly to bee stings when they were younger. They required adrenaline and an ambulance trip to the hospital. Thankfully, now they do not have adverse reactions. Maybe their maturing immune system has been able to cope with the infrequent bee venom incidents.

With children I think it is all about education. If the grass has flowering clover they need to wear shoes, if a bee lands on them they just stand still and let it fly away without running or flapping their arms in terror. Ultimately you, as a parent, need to make the call on whether bees are a good idea in your backyard if you have young children.

Avoiding stings
Bees can pick up on your mood so it is always imperative to open your hive in a calm and controlled manner. If you manipulate the frames gently and calmly, the bees will most often go about their work unperturbed. Avoid any jarring or bumping of the frames. Never bang the roof, as bees are very sensitive to vibrations. If bees crawl on your hands or arms do not swat them away; leave them, as they are unlikely to sting. Always work your hive from the back

With supervision and support older children can manage their own hives. Bees are a wonderful learning tool and create environmentally aware citizens.

or side so you do not block the entrance. Use a smoker very lightly and always check that the smoke you are blowing is cool. If you use excessive smoke it can have the opposite effect and anger the bees.

Never swat at a bee. If a bee gets caught in your car, lower the window to allow them to escape or turn on the fan or air-conditioning so they are pushed to the back of the car.

Sometimes you may open your most placid of hives and find some very angry girls. Bees will show their displeasure by standing with their tails aimed at you and their stings extended. If your bees buzz your head or start bouncing off your veil, it is time to quickly (but calmly) close up the hive and leave your visit for another day. Sometimes the weather can make bees more defensive. An approaching thunderstorm is thought to make bees more defensive.

Many people will tell you that dark colors need to be avoided, as bees are more likely to sting. I have never found this to be true. I habitually tend my hives in jeans and dark colored clothing without incident. Lighter colored clothing certainly makes it more comfortable in the heat of summer. Strong perfume should be avoided. Bees also are very sensitive to carbon dioxide so avoid breathing or blowing onto the frames. I do avoid wearing woolen jerseys as a bee can get easily tangled in the fiber and be forced to defensively sting. For the same reason, I usually choose to wear a veil, as my thick, curly hair seems to be a magnet for bees, which burrow in my hair and then often sting.

Time commitment

Bees do not take a large amount of time when you are keeping them as a hobby in your backyard. If you can spare half an hour every fortnight to manage your hive, hobby beekeeping is for you. When you first start your new hobby, things will take longer, as you struggle into your bee suit, toil over a smoker which does not want to remain lit, and pluck up the courage to open your hive. After time all these things will become second nature to you. Very soon you will be able to conduct a hive check in a matter of 15 minutes. Half of that time is spent struggling with your bee suit!

Often a hive check will just entail opening the roof and peering in to look down the combs, or at the end of the combs if it is a Top Bar hive, to ensure that the bees have ample room. You may perhaps remove one or two frames to look for evidence of food stores and a laying queen. For a quick hive check this is all that is needed; you don't have to go through every frame every time you open a hive. This check can take a few minutes, with some practice.

Keeping a beehive does not require the time it takes to keep a pet dog. The bonus is you can still go away for extended lengths of time without having to pay for a house sitter or expensive kennel fees.

Beekeeping duties vary with the seasons. Early spring through to mid-summer is the busiest time, with fortnightly visits to the hive a minimum. Over autumn a monthly visit will suffice and over the quiet, almost dormant, winter months you can leave your hive to slumber in its warm cluster inside the bee hive.

Any extended holidays are best taken during late autumn and winter when the hive requires less management.

A rough guide to time commitment for a backyard hive

EARLY SPRING	LATE SPRING	SUMMER	AUTUMN	WINTER
Visit the hive every 14 days, on warm, still days, to ensure the queen is laying and there are adequate food stores.	Visit the hive every 10–12 days to check for signs of swarming. Look for excessive drones and queen cells.	Visit the hive every 3 weeks to check that the colony has space for the expanding bees and honey stores.	Monthly visits to check on harvestable honey and ample honey and pollen stores for winter. Check that wasps or robbing bees are not evident.	Observe the entrance of the hive for flying bees on a warm day. Open hive once in late winter to check that the bees have food stores.

What does it take to be a great beekeeper?
"If you are not making mistakes you are not learning."

Inner strength
When you start beekeeping, not everything will go to plan. It can't when you are working with insects. The stark fact is you will lose hives as your colony dies through disease or some other cause. When this happens—and notice I say when, not if—you need to pick yourself up and try and analyze what went wrong and try not to repeat the mistake. Beekeeping is always about trial and error. This is the reason why I recommend that you start with at least two hives. This way you still have one surviving hive, which you can split when the time is right.

Critical thinking
There is no set formula you can follow. Beekeeping is not a hobby that suits a paint-by-numbers mentality. Seek out information from books, the internet and others who are more experienced, but always evaluate the information before you load it into your memory. As a backyard beekeeper, always question commercial beekeeping practices, as their desired outcomes are very different from ours.

Learn how colonies choose to live in the wild and try and emulate that in your own beekeeping practice. An independent thinker will be the best kind of beekeeper, as you will always be questioning and making informed decisions.

Be nosy and probing
As a beginner beekeeper, to learn, you need to open your hive regularly. Be curious and watch, assess and learn every time you are in your hive. Keep hive notes and document what you see at each hive visit so you can compare. If you see something new or interesting research it online or look it up in books so you can gain a deeper understanding; the bees will never fail to teach you something new. On that note, never trust a beekeeper who states they know everything they need to know about bees—this means they have stopped learning.

Show respect
Accept that the honey bee will always know better than the beekeeper concerning anything that goes on within the hive. The best beekeepers are the

bees themselves. Before you do any sort of manipulation of frames or boxes, stop to consider if the bees would choose to do this if they were a wild hive.

Show calm under fire

You will get stung and how you react is paramount. Being calm and considered in everything you do within your hive will help prevent stings, as the bees are more likely to remain calm. If you are a person who gets easily anxious or fearful during a stressful episode, try and work through these incidents in your mind before they happen and have a plan for what you would do. If you have role-played it in your mind, you are more likely to have the skills to negotiate your way through it.

How hard is it to keep bees?

There is certainly a lot to learn about keeping bees and in this modern world where keeping them alive, let alone thriving, is becoming more and more difficult, it is a challenge. Your first year of beekeeping will be a steep learning curve but I can guarantee it will be very satisfying. You will never feel bored with your new hobby. The beauty of backyard beekeeping is that you can be guided by the bees, as there is no pressure to produce copious amounts of honey to keep a commercial beekeeping operation afloat.

What about children and pets?

I have all manner of feathered and furred animals in my garden and have had no problems with the interface between bees and animals! My ducks and chickens pay no attention to the Top Bar hive that is situated in their permanent run. My labrador, as a puppy, had one sting on her nose and quickly learnt that those buzzy things deserved some respect and distance. The cat can often be found lounging in the sun on top of the roof of one of my hives.

Children seem to have an inborn interest and adoration for bees. With a little education they can learn how to behave around flying bees and that a bee only stings if it is stomped on or threatened. My kids are now very calm and accepting of bees in our garden. Sure there is the odd sting, but that is really just because there is now a population of pollinators in our garden. Before our hives there were no bees. When the lawn has flowering clover or dandelions (yes, my lawn is no bowling green!) they know to wear shoes.

CHAPTER 2

GETTING STARTED

Do your research

I recommend turning yourself into an information sponge before you get your bees. Read every book you can gather, do some online research and perhaps attend a meeting at your local beekeeping club, if you have one in your area.

When you first start to research what is involved and what there is to know about beekeeping, it can seem very daunting. It is easy to be put off with all the bee jargon out there. Combine this with the confusing fact that when you ask two different beekeepers the same question you get two completely different answers.

Many new beekeepers actually feel overwhelmed with all this information they think they need to know. You are never going to know everything there is to know about bees, only the bees themselves know that!

However, do not take as gospel everything you read and hear. Beekeeping is very local and what may work for a beekeeper in one part of the world may not work for you in your local climate. This is because of differences in climate, flora, rules and regulations, wildlife and environmental issues.

Beekeeping clubs

Beekeeping clubs can be a font of information and advice, but tread carefully in your initial interactions with club members. Bee clubs are changing, but there are still some older members who are retired commercial beekeepers. They have known only one way of beekeeping, that of the commercial Langstroth hive. Unfortunately, some are very averse to any new ideas or hive designs and may try to encourage you to follow their way of beekeeping. I personally think that a backyard beekeeper should not try to turn themself into a mini-commercial beekeeper. We have a huge potential to look at bees in a more bee-focused light and create a flotilla of healthy, unstressed bee-friendly colonies in our gardens.

Developing your own beekeeping ethos

The most important thing is to listen and learn from an array of beekeepers and then take what you want from them. It is important that over these initial months you discover your own ethics and beliefs about how you want to keep bees and I hope that after reading this book it will be organically and as bee-focused as possible, always putting the bees first rather than honey production.

The way a commercial beekeeper goes about operating their hives can be

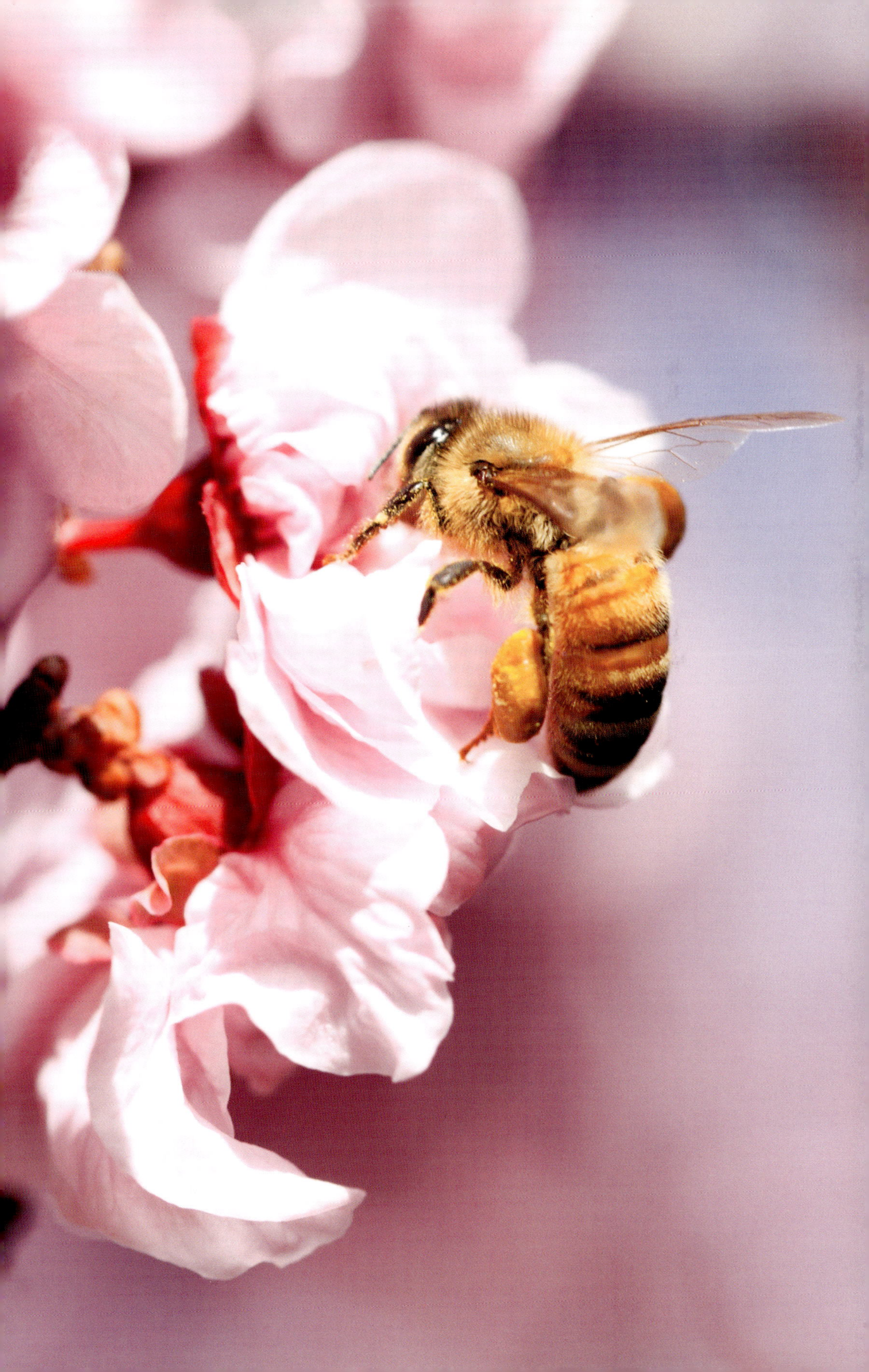

very different to how a hobbyist beekeeper manages their one or two hives. As a hobbyist beekeeper you will be in a unique and privileged position. A commercial beekeeper is managing thousands of hives, they have a budget to work to and huge costs in machinery, labor, fuel, legislation fees and processing plants that they have to pay for. Their hives are treated as units which must have a return on investment. Honey is money and the bees are forced to be as productive as possible to make the business profitable, often at the detriment of the hives' health.

A backyard beekeeper has more time, can experiment with more natural and bee-focused ways of sustaining bees and can be more focused on providing what is best for the bees rather than producing the most honey they can from the hive each year.

Space requirements

Beekeeping is a great way for urban people to reconnect with nature as this hobby does not require large amounts of room, a garden, or even a backyard. As bees fly to neighboring food sources, it is not critical to have your own backyard full of lovely blooms. A bee will forage comfortably in a 1.8 mile (3 km) radius from her hive. In times when nectar sources are scarce she can fly up to 3.7 miles (6 km) away from her hive to collect nectar or pollen.

I have successfully kept a Top Bar hive on my small upstairs patio, just off the kitchen, for a number of years. In fact, a beehive that is raised a few stories from the ground decreases the chance of people being stung if they walk through the bees' flight path.

A Top Bar hive requires a footprint of around 23½ in x 4 feet (60 cm x 1.2 m) with space at the rear for you to stand while you are working your hive. Warré, Flow and Langstroth hives require around 23½ in (60 cm) squared with space at the rear or side to stand. Bees are being kept successfully on rooftops and balconies in large cities around the world.

Legal requirements

Every city, council or city hall will have different rules and regulations on where and even if you are allowed to keep bees in your backyard. Every beekeeper must follow the regulations in their own area. It is important to follow the local rules and regulations before setting up a backyard hive as you want to avoid breaking any laws and paying fines. And there is also a risk of losing your hive and undoing all your good work.

The best process to follow is to search online for beekpeeing in your area, or inquire at your local council or city hall. Your local beekeeping club may also be able to point you in the right direction. It is common in many countries that beekeepers are required to be registered and the location of their apiaries must be registered. Often there is an apiary database. Once registered you may be assigned a beekeeper registration number, which often needs to be shown at your apiary site. I pay a yearly fee per apiary site and I have to complete my paperwork annually. This will differ for each country.

Most often the requirements and laws focus on preventing a nuisance to your neighbors. If you can win over your neighbors you're more likely to be successful in getting approval. Tempt them with the idea of free pollination, free pots of their garden honey (your bees will be visiting their garden) and the fact that they are doing their bit for the environment without even trying!

As urban beekeeping is gaining popularity some bylaws are lagging behind. In the past there has been little necessity to work urban beekeeping into any local regulations or bylaws. Some regulatory authorities ban beekeeping outright in urban areas, but luckily this is becoming less common. If you live with this sort of legislation the only way forward is often to change policy. This is luckily happening with regular occurrence around the world. Just recently Los Angeles made urban beekeeping legal after protracted lobbying by local beekeepers. If your region has no laws to support urban beekeeping why not be a policy changer?

Getting bees

This is perhaps the most difficult aspect on your path to becoming a beekeeper. You can obtain bees any time of the year but spring and summer is the best time as more bees are available. This is because hives are being split, queen bees are being reared and swarms are being cast off from strong hives.

If you know of a local beekeeper, they may be in a position to split a strong hive and be able to sell or give you a nucleus colony. A nucleus colony is a mini colony normally with four to five frames of comb, brood, honey and pollen and a laying queen. This "nuc" can then be placed into a hive and get on with the job of expanding. If you have joined a beekeeping club there may be members through the warmer months who have nucleuses available. The trick with a nuc is that it must be on the same size and design of frame as the hive you are starting out with so they can be easily transferred over. For example, a Warré hive frame will be too small to fit into a full depth Langstroth super.

Swarms are a great way to start your beekeeping adventure. Catching a swarm is a wonderful experience to share with children, who often have a natural affinity with and interest in bees.

A nuc on full-depth frames will not fit into a three-quarter sized Langstroth hive. Remember, like with like makes it easy.

Catching a swarm is the free way to obtain a colony of bees. When a hive becomes so strong that the colony becomes cramped, the old queen will swarm from the hive with most of the mature foraging bees to find a new home. Between leaving the old hive and finding a new home is the optimal time to gently retrieve the swarm and settle them into your new hive. I never tire of catching a swarm, it really is exciting, with the smell of the honey-gorged bees and the sound of thousands of humming wings.

If you live in the United States, buying package bees is another option but one that is generally not recommended. Try and obtain bees the aforementioned ways if possible. A package of bees is a wooden screened box, which has around 3 pounds (1.3 kg) of unrelated bees in it, complete with a caged queen. This mass of bees is introduced into your hive and it is hoped that everyone settles down, makes a big happy family and gets to work building new comb. Often this is not the case. The queen may not be accepted or the bees may abscond and leave the hive.

See Chapter 7 for more details on how to get bees into your hive.

CHAPTER 3

THE LIFE OF BEES

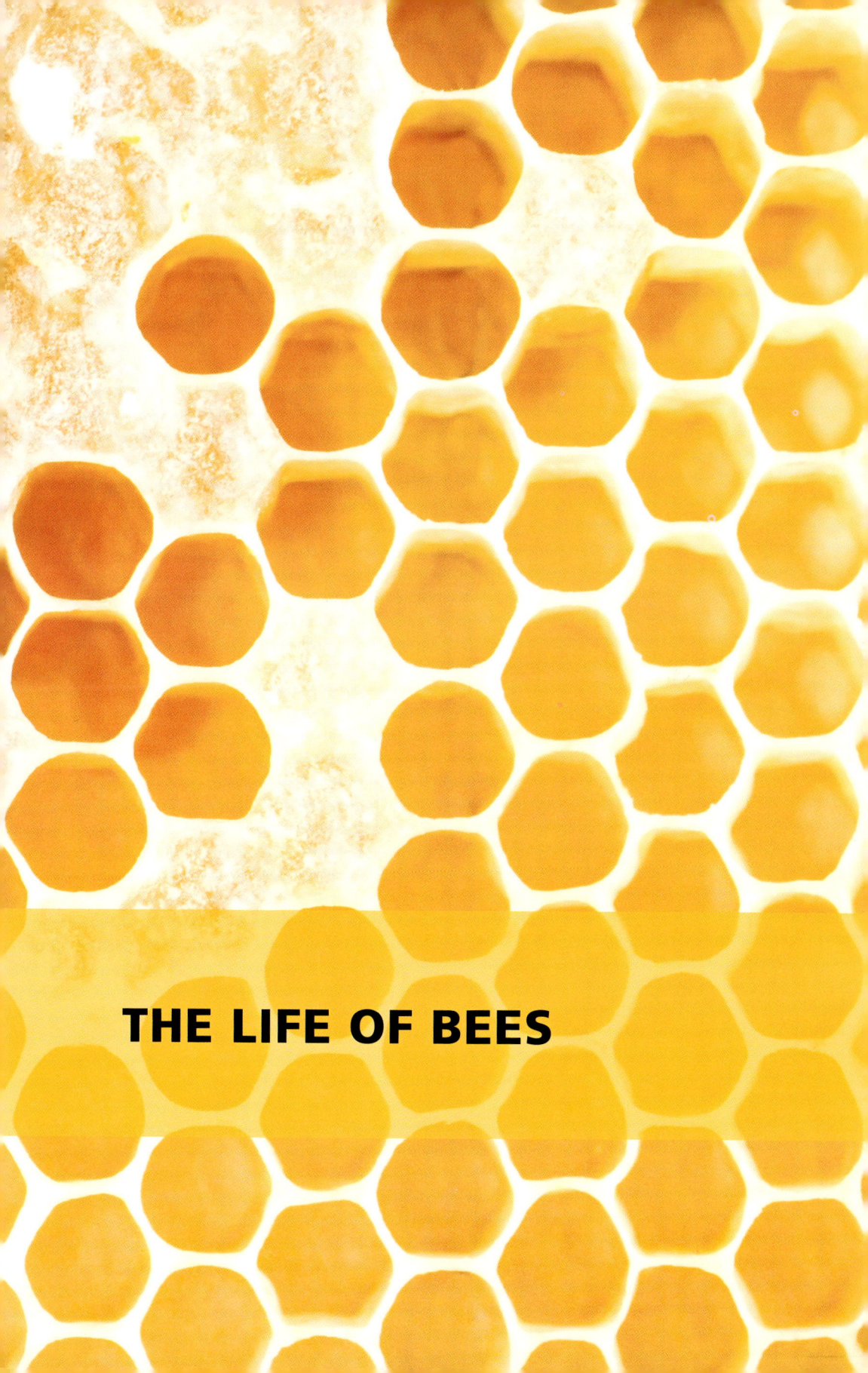

To be a competent and considered beekeeper you need to understand the "why" not just the "how". Many new beekeepers just do a manipulation in their hive because this is what they have been told or shown by a mentor or other beekeeper. This is often done without really understanding why. If you gain an in-depth understanding of how the honey bee functions then you can be more informed and mindful of why you are doing anything concerning your bees and your hive. Understanding how a hive functions will make you think about which practices are unnecessary if you want to keep your bees in a more natural and bee-focused way. Many beekeepers, I think, have a "paint by numbers" mentality when they are beekeeping. I want you to learn to be your own beekeeper, not someone else's copycat.

The honey bee world

The all-encompassing desire of a honey bee colony is to reproduce and survive through winter. Bees build a social structure and wax comb to achieve this.

A bee colony contains three adult and three developmental stages of honey bee. The three types of bees share some fundamental characteristics. The bee is made up of three distinct body parts: the head, the thorax and the abdomen. They have six legs, four wings, two compound eyes, three simple eyes and two antennae.

The queen

In a normal hive there is one queen, but it is not uncommon to have two related queens. Often this will consist of a young, newly hatched daughter of the established queen. The colony allows these two queens to coexist beside each other. Often beekeepers are not aware they have a two-queen colony, as once the first queen is spotted the beekeeper moves onto another task in the hive. A laying queen has mated with unrelated male bees (drones) soon after she hatched and is the only reproductive bee in the colony. The queen is the mother of the colony and is essential to its survival. Without a laying queen the colony cannot replace worker bees and over time will die out.

A queen starts her life as any other freshly laid fertilized egg in a cell. The colony determine if the old queen needs replacing. They do this by sensing decreased amounts of queen pheromones. Worker bees will choose a freshly laid egg, less than three and a half days old and the instant it turns into larvae it is fed copious amounts of royal jelly. All larvae are fed this special substance for the first three and a half days but then the worker and drone larvae diet is switched to pollen. The egg deemed to be the new queen is continued on the royal diet of jelly. The worker bees then begin building an elongated, peanut-shaped wax cell to give the queen space to develop.

On the sixteenth day a developing new queen will emerge. She immediately commences a seek and destroy mission. She will search out any other queen cells. If found, she will rip them open and sting to death her emerging sisters, so only she reigns supreme. If you find numerous open queen cells in your hive, look at them closely. If one has a neat little opened lid at the end of the queen cell, this is likely to be the new queen's cells. If any have an opening from the side, then these are the queens that have been ripped from their slumber and stung to death by their sister and dropped on the hive floor.

A queen has a stinger but it is not barbed like a worker bee. This means she can sting numerous times but luckily she will only normally use this to kill other queens.

After the new queen emerges she spends the first week of her life inside the hive. She is treated like any other bee and must beg for food from the returning foragers. On a clear, still, sunny day the new queen will leave the hive in the afternoon and go on a mating flight, which is generally fairly short, at under an hour. She heads to drone congregation areas, normally around 1.8 miles (3 kilometers) away. This distance increases the chance of new genetics being introduced to the colony. These aerial mating zones are a bit like street corners where drones are hanging out looking for virgin queens to mate with. Drones are attracted to her by a sex attractant pheromone she emits.

Mating is done high in the air. When she mates with a drone its sex apparatus is ripped from its body in the act. The drone's genitalia can sometimes be seen protruding from the queen as she returns to the hive from her mating flight. She will mate with eight or more drones. She often will take several mating flights over several days. Each time a queen mates the drone's semen is deposited in a layer inside the queen. Layers build as she completes her mating flights. As the queen starts laying she uses one layer at a time. This is why sometimes a hive can suddenly become more aggressive. This change in behavior can indicate that the queen mated with a drone with aggressive traits and this is the semen she is using to fertilize eggs.

After two to four days back in the hive she is fully fertile and ready to start egg laying. Now that she has mated she starts to emit a distinctive queen odor which is passed through the colony. This smell tells the colony that the queen is present, the bees are in the right hive and all is well. Now that the queen is mated she takes on her special royal status and is treated with reverence by all in the colony. Worker bees look after the queen, feeding her, cleaning and grooming her and taking care of her every need.

A queen alternates resting with egg laying throughout a 24-hour period. During the busy spring and summer times the queen will alternate egg laying for 10 minutes with resting for 10 minutes. A queen can lay over 200,000 eggs a year!

Over a queen's lifetime she may lay for several years. In today's modern age of beekeeping, with many beekeepers forced to use toxic chemicals, we are finding that a queen may only last a year or less before she has to be replaced by the colony. A queen can lay up to 1,500–2,000 eggs each day during the warmer months. During the colder winter months, depending on the temperature, she will decrease laying or stop all together for a time.

Many people think that it is the queen who is the leader of the colony and dictates what the colony will do but it is actually the worker bees who run the colony. The queen is really nothing but an egg-laying machine and at the mercy of the worker bees. The worker bees are constantly judging the queen's laying ability by assessing the scent or pheromones she gives off. When her scent begins to weaken through old age, insufficient number of drones she has mated with, or ill health, the workers will start making preparations to replace her with a new queen.

The queen emits a scent that all workers can detect. This scent is passed around the hive through the bees rubbing against each other. This pheromone is heavy so does not travel through the air but rather from bee body to bee body. If the queen is removed or dies, it only takes two hours for the hive to register the queen's absence and they will start doing everything they can to replace her.

Worker bees

Worker bees are all female bees with functional ovaries so it is possible for them to lay eggs but, as they have never mated, they will always only lay unfertilized eggs resulting in drones. Worker bees also possess a sting and pollen sacs on their back legs. The interaction of queen pheromones and brood pheromones within a normal functioning hive suppresses the activation of these worker bees' ovaries and their desire to lay eggs. Worker bees are the most abundant bee in the hive. In the height of summer there can be up to 60,000 worker bees making up the colony.

The worker bee is completely responsible for all the discussions and day-to-day running of the hive. They execute every duty inside and outside the hive. Depending on their age, they have set duties to perform and they learn these duties by the pheromones given off by the rest of the hive—by other workers, the queen and the developing brood. After emerging, they feed on pollen stores and stay within the brood area as they are very sensitive to any light. As pheromone levels rise they start the in-house duties of a house bee. They clean the brood cells between a bee emerging and a queen laying a new egg. Soon a specialized gland in their head begins to function which enables them to produce royal jelly. When this gland is mature the worker bees starts her life as a nurse bee. She will produce royal jelly and feed it to the youngest larvae and to any developing queen larvae if there are any.

A bee's next duty as she ages is nectar ripening. During this duty a worker bee will position herself near the entrance. She greets incoming foraging bees, who have stomachs filled with foraged nectar. Transferring this nectar into her stomach, she takes it to an area above the brood area or to the end of the hive and starts to mix the nectar with her stomach enzymes, which aids in ripening the nectar into honey. She will then deposit this tiny drop into a cell. When this cell is filled to the top and the excess water content is evaporated off by the bees fanning it with their wings, it is capped with a layer of thin wax.

Other similar-aged worker bees help to store pollen. Collected pollen is packed into cells. The bees use their head to punch the pollen down and compress it. Once a cell is three-quarters filled, a very thin layer of honey is used to seal the cell. The pollen will ferment and turns into bee bread, which is fed to developing worker and drone larvae as a high-protein meal.

When you view stored pollen you will notice that the pollen only ever reaches three-quarters of the way to the top of the cell. This is completely normal. The pollen may be many colors ranging from white to yellow, orange and even purple depending on the species of plant it was collected from.

When a worker bee is around 12 days old she becomes capable of producing tiny flakes of wax from a specialized gland on the lower side of her abdomen. The wax is used to build and repair comb and cap honey-filled cells. These wax scales are translucent and white. As a wax flake is produced it is transferred to a worker bee's mouth where it is chewed to make it soft and pliable. A worker bee has the ability to produce wax until she is 18 days old.

At 18 days old, a worker bee who has lived her entire short life in the warm buzzing comfort of her dark hive becomes attracted to light as her hormone

This field bee has the task of collecting pollen. Her body is charged so that pollen is attracted to her hair. As she collects it she will store it in pollen bags on her hind legs.

levels change. This change leads to the next and final stages of her life. The desire to explore the light which is shining through the hive entrance holes urges her to explore further.

Between the ages of 18 and 21 days, the worker bees become guard bees at the front of the hive. They protect the hive from attack by wasps, other bees and all manner of animals. Every bee who enters the hive is smelt by the guard bees. If a bee does not have the distinctive odor of the hive it will be challenged and driven away or permitted to enter if she comes with a gift of nectar. Flying bees that have become separated from their hives or lost can often coax their way into a new hive home by coming bearing gifts of pollen or nectar. The guard bees are on duty 24/7. Guard bees are the ones who will rush at you if you stand too close to the hive entrance. When you open the hive and the bees are not in a convivial mood, the guard bees are the ones that will sacrifice their lives by using their sting to protect the colony.

On warm, sunny days in the early afternoon it is normal to observe a rush of activity around the entrance of the hive. When you look closely you can see many worker bees flying in a figure of eight pattern facing the entrance of the hive. These are young field bees learning their forager groove. Initially a new forager bee will only make short flights from her hive but as her inbuilt navigator system becomes more developed and she learns to recognize the local landmarks near her hive, her flights become longer. Field bees have the job of collecting pollen, nectar, water and propolis—all important elements for survival.

A worker bee will live for around six weeks over the busy warmer seasons. She will most likely die out in the field by being eaten by a bird or not being able to make it back to the hive due to a sudden change of weather. If you look closely at the worker bees in a hive you can pick out the older bees. Older bees' wings may have small rips along their margins, the thorax appears smooth as all her fuzzy hair has been worn off through the millions of trips into the center of flowers and the countless times she has bumped into her sisters within the hive during her hive communications. Bees most commonly will die out in the field but if they do die within the hive they are removed from the hive by undertaker bees. Their body is carried away from the hive and dropped.

Drones

In a commercial hive, drones get a very unfair rap. They are thought of as useless, lazy, food-hogging layabouts. This is because drones do not collect nectar, nor do they perform any hive duties.

Drones are reared by the colony only during certain seasons and conditions. Only a strong, healthy hive will produce drones due to the added drain on precious resources. Drone rearing begins in spring when food resources become reliable. A drone egg is produced from an unfertilized egg laid by the queen. Drones take the longest time to develop from an egg to an adult bee—24 days. They only live for around two months. Drones do not possess a stinger. They are the largest bee in the hive. They have huge eyes, all the better for spying virgin bees to mate with, and strong wing muscles for chasing after her. Their main objective in life is to mate with a virgin queen from another colony to pass on the genetics of their mother, the queen.

It is hard not to like drones. They are big and bumbling. Covered with fuzz and owning no stinger, they are the teddy bears of the bee world. They are sex-crazed fanatics, though of course in the nicest way!

This natural comb shows worker brood and drone brood. The drone brood is larger with thicker walls and is normally found on the edges of comb. A queen will only lay drone eggs in these larger cells and in times of high nectar flow the hive will use empty drone comb to store honey.

Every sunny afternoon all mature drones (older than 12 days) start their jumbo jet engines and fly off looking for some nooky in the sky. It is easy to hear a drone as their huge wing muscles and large body make them sound like a jumbo jet rather than the fighter plane sound of the worker bee. Within a few miles of his home hive, drones will meet in a drone congregation area. These are areas used season after season as a meeting place for drones and virgin queens. If a virgin queen happens to fly into this area it is all on! Drones will chase the queen and the fastest drone will make contact with the queen high in the air. After mating in mid-air, the lucky drone suddenly becomes ill-fated as he falls to the ground with his penis ripped out of him by the queen. We can only hope he dies with a smile on his face knowing that his job has now been done.

In a hive that uses foundation frames it is common to see very few drones as all the cells of the foundation only allow worker brood to develop. A colony will try and build drone comb in any nook or cranny in an attempt to raise

It is important to learn to recognize drones in your hive. They are larger than the worker bees and have very large eyes. Large eyes make it easier for a drone to spy a virgin queen high in the sky. They have a blunter-shaped end than the worker bees and possess no stinger. PHOTO: *Alphapix.*

drones. In a hive which allows the colony to build their own natural comb, it is normal to observe a population of drones making up around 15–20 percent of all adult bees. This simple fact is telling us that it is imperative for a healthy, natural hive to have a good population of drones. As backyard beekeepers we should allow our colonies to raise this natural amount of drones. Drones help with the social cohesion of the colony. I believe that drones are important for the moral and innate management of a well-functioning hive. Drones are also important for your local area as they keep the area stocked with genetics for breeding. Drones are fundamentally central to their colony's success in the eternal evolutionary competition to pass genes on to future generations.

The mechanisms of a hive

Bees don't just make up a hive, rather it is a carefully balanced and crafted array of unique compounds which all work to provide an optimum environment for the bees to live, work and prosper. As a beekeeper it is our

This ball of propolis scraped from the hive is valuable in creams and as a mouthwash due to its natural antiviral and antibacterial properties. Many beekeepers suck on a piece as they work the hives.

responsibility to recognize and help maintain these elements. A single colony requires 44 pounds (20 kilograms) of pollen as a protein source, 264 pounds (120 kilograms) of nectar as an energy source and 845 fl oz (25 liters) of water for cooling the hive through the season.

Propolis
Propolis is a sticky resin-like substance which bees collect from certain species of trees. Trees produce this resin to protect emerging buds from fungal and bacterial diseases. Propolis is a complex mixture of tree resins, digestive bee enzymes, essential oils and pollen. The bees use it as a glue. Bees can only collect propolis when temperatures are above 64°F (18°C). In colder temperatures the propolis is too hard to collect.

The bees use propolis as a form of glue and filler in their hive. They will reduce a too-large entrance with propolis, making their home easier to defend. They will stop up any gaps and cracks to keep out cold drafts. If a foreign animal, such as a mouse, enters the hive and is stung to death the bees will entomb it in propolis to protect the hive from decaying bacteria.

In the past bees have been bred to reduce the amount of propolis collected to make the beekeeper's job easier when inspecting the hive. A hive which is glued shut with propolis makes for lots of prizing and wrenching with a hive tool. Propolis always coats gloves, hive tools and other beekeeping equipment. Only recently have beekeepers appreciated the benefit of propolis. Recent studies have discovered that bees do not have an internally developed immune system but rather the propolis acts as their external immune system. Propolis has amazing antibacterial, antifungal and antiviral properties. Collected propolis can be used in tinctures and research has shown that it is especially beneficial in the treatment of gum disease.

The bees will coat the interior walls of the hive with propolis for hygienic reasons and also to create traction to make it easier for them to walk up the sides and to anchor new comb to. Bees will fill any cracks in top bars of a Top Bar hive with propolis to contain the important temperature and hive scent of a hive. I always find it easy to judge the temperature of a day with my Top Bar hive. When I separate two top bars and there is a crack or a pop I know that the propolis is cold and brittle and the combs will be more robust. If I separate the top bars and the propolis stretches like mozzarella cheese, the temperature is hot and I must be very careful with soft fragile wax combs.

Wax comb

Comb makes up the largest part of a hive and is at the heart of any healthy hive. The comb of the hive works like the inner organ of the hive. Wax flakes are produced and secreted from a specialized gland on the underside of a worker bee's abdomen. In normal situations only worker bees between the ages of 12 and 18 days have the ability to make comb wax.

Comb is the food storage system, the nursery, the womb, the communication organ, the colony's immune system and the hive's insulating system.

A bee will spend around 90 percent of its life on comb. All bees are raised inside comb cells.

Comb is covered with a very thin layer of propolis. If you look closely at new comb you may observe orange-tinged edges of the comb cells. The propolis, which the bees have coated the new comb with, acts as an immune system. Bees also coat the inside of each cell with propolis before the queen lays her next egg in it.

When a bee does her waggle dance to tell her sisters where to forage for

A thin layer of propolis can be seen along the bottom edge of this frame. It has been used as a seal to stop heat loss and invasion by insects such as ants or wasps. PHOTO: *Alphapix.*

nectar, the vibrations travel easily through naturally built comb. This allows these messages to be transmitted to other bees without having to be near to the messenger. Who knows if this communication is possible when beekeepers choose to use the ridged and unnatural plastic foundation?

Comb is also the food larder to ensure that the colony survives through the lean months of winter. When a cell is filled to the top with nectar and has been suitability evaporated to turn it into honey, a thin layer of wax is used to cap the cell. This is called honey, and is the only food in the world that will never decay or spoil. Edible honey has been found in tombs inside the pyramids of Egypt.

Each comb or frame of comb within a hive has an important and significant role within the hive. It is important for a beekeeper to respect this and to not switch around the orientation and order of frames within the brood nest. It really is akin to moving one of your vital organs from inside your torso to place it on the end of your foot.

Bees always, without exception, build wax from the top down. As comb is filled, the colony moves downwards. Bees prefer to build their wax home in a round or oval shape. If you are lucky enough to ever see a wild colony you will notice that the comb is not in the shape of a neat square box but rather round or oval. This round form is much more conducive to regulating warmth and humidity as corners trap cold air.

A dark enclosed space is needed for efficient wax secretion from the worker wax glands. Bees in darker hives secrete more wax and build more comb than bees exposed to light. This is a fact worth remembering if your hive has a viewing window. Ensure that the cover remains in place when you are not looking into the hive. Whilst comb building is taking place, safeguard this important bee function by keeping hive openings to an absolute minimum.

Regardless of hive design, the most important thing you can do as a backyard beekeeper is to ditch the plastic and wax foundation. Allow your bees to build their own precious comb to their own unique needs. Foundation is a modern beekeeping method which is all about making life easier for the beekeeper, and more profitable in terms of honey produced. This is all at the detriment of our honey bee. Allowing your bees to build their own comb allows the bees to live how nature intended and provides an efficient and operational home.

The magic that is wax
Naturally drawn wax is like the uterus of the hive. It is precious, unique, highly crafted and engineered and a precious resource. Beeswax is produced by honey bees and used in comb construction, comb repair and capping the cells when they are full of ripe honey or developing larvae. The wax is produced by individual bees on their lower abdomen and exits their bodies as thin, small flakes of wax.

Worker bees producing wax need to eat large amounts of honey as a source of energy for this taxing role within the hive. To develop well-functioning wax glands worker bees also have to eat pollen during the first five or six days of their life. Wax is secreted primarily during warm weather when foraging and a nectar flow is active. Bees are unable to build comb during the very cold months or if there are no food reserves in the hive or no nectar flow in their area. Workers will gorge themselves with honey and hang under the bars or to the side of comb secreting these wax flakes. This is called festooning or daisy chaining. Drones and queens do not have this abdominal wax gland.

When a bee produces a flake of translucent wax another hive bee will remove it from her abdomen and chew it to make it soft and pliable and then manipulate it with her jaws and legs to build the comb.

The hexagon-shaped comb, which the bees construct to precise dimensions, is the most efficient shape whilst using the smallest possible amount of wax to store the highest volume of honey or pollen. It has been shown to be the strongest possible shape while using the least amount of material.

When assessing your hive in the spring and summer months you can assess new comb building by witnessing the festooning of bees off frames as you remove them from the hive. On a bottom board you will also regularly see tiny, white flakes of wax which have fallen from the builder bees.

Freshly built wax is always white and very pliable and fragile. As it ages it becomes more yellow, leading to orange and then a darker brown if it has been used for brood rearing. The darkening color is because with each round of brood rearing the developing bee leaves behind a small amount of bee waste in the form of feces and the cocoon lining. Experiments have shown that the survival of eggs and larvae is higher in cells that have been preciously used compared to survival in new cells. As brood comb goes through many cycles of brood rearing it has the potential to harbor disease though.

Very old comb is nearly black and has lost all transparency. Old brood comb can harbor disease and it is a good practice to regularly cycle old comb out of the hive and replace with organic foundation or allow the bees to rebuild the empty frames themselves. An easy way to cycle out older comb frames is to mark and place these frames at the end of the colony (or above the colony in a Langstroth), which encourages the colony to use these fames for honey storage. At the end of summer you can harvest the combs as honeycombs. Once the honey has been spun out of these combs they can then be melted in a solar wax melter.

You can number the frames to keep track of the age of comb and aim to recycle the comb out of the hive every three to four years.

Not all comb is the same

When you look closely at a naturally drawn comb you may notice that some of the cells are larger than others. Often these cells can be found near the bottom edges of the frame of comb. This is drone cell comb. As the drone bee (male bee) is larger he requires a larger-sized cell in which to develop. Bees will draw these larger-sized cells around the bottom edges of the frame or often build an entire frame of brood comb. It is thought that the colony chooses to build

A swarm has started building a natural nest in this coir basket. A swarm of bees can build their own natural comb incredibly fast. These combs were built by this swarm over a ten-day period. A beekeeper is carefully cutting out these combs and transferring the colony and queen into a hive box.

the drone comb around the lower edges of a frame as this is the area of a hive which is most prone to chilling or temperature variations, which can lead to brood loss. The hive is willing to sacrifice drone brood rather than worker brood.

A queen will only lay an unfertilized egg into a drone-sized comb. This unfertilized egg will only develop into a drone bee as worker bees develop from a fertilized egg and the queen can ingeniously lay these different eggs at will. Before laying an egg in a cell the queen will measure the diameter with her front legs. If she judges the opening to be large she will lay an unfertilized egg

This wild hive has built natural comb. All bees will maintain a bee space between each comb so that the bees can reach each side of the comb and attend to the colony's needs.

which will develop into a drone. Only fertilized eggs are laid into the smaller worker cells.

Drone cells can also be used by the colony to store honey.

In a hive which uses foundation comb the bees do not have the choice of building their own comb. Foundation is made by a machine. The raised hexagons embossed on the wax sheets are all one size and follows the mantra "one size fits all". Foundation is recycled wax (and now more often plastic) that has been pressed into thin sheets with the shape of the worker cells embedded into the wax or plastic. In a hive that uses this system, the queen is forced to lay only worker brood. Wax or plastic foundation frames are used to maximize honey production. A hive which only produces worker brood and has man-made cells in which to lay cannot build their own nest, but honey production is maximized as the colony's efforts go into collection and storage of honey and only raising worker bees. This maximizes honey production but what is it doing to the health of the colony?

Added to this "one size fits all" problem is someone's past "brilliant" idea to make the foundation cells a bit bigger than what a colony would make in a natural hive. The thinking behind this is that worker bees will become larger, thus be able to fly longer distances and harvest more honey and then of course the larger cells can hold more honey for the beekeeper to reap at the end of the season.

The difference between natural cell size (0.19 in/4.9 mm) and standard cell size (0.21 in/5.4 mm) is considerable when you are dealing with a small insect. This larger cell size makes it easier for the varroa mite to breed and multiply in a managed hive. Enlarging the cell size causes the development time of the bee larvae to extend up to 24 hours. This provides the varroa mite extra time to breed, develop and feast off the developing and helpless bee larvae.

A wild hive and a natural comb hive will have on average 15 percent of the population as drone bees and 15 percent of the brood nest cells are drone-sized. When a colony on foundation is given some space to build their own natural comb they instantly start building drone comb in order to regain a population of drones. Drones are important for the moral and intrinsic health and order of the colony. Drones carry the genetic constitution of the queen and her colony.

Backyard beekeepers should see drones as providing peace and serenity to the hive rather than the commercial beekeeping attitude of drones being freeloaders, idlers and spongers.

Allowing your bees to build natural comb, regardless of hive design is, I think, the most important practice for beekeeping. PHOTO: *Alphapix.*

Should you use foundation wax sheets or allow the bees to build their unique comb?

Foundation is thin sheets of recycled beeswax, which have been rolled and pressed with a hexagonal pattern of comb cells.

Reasons **for** using foundation:
- Encourages the bees to draw out comb in the size and orientation the beekeeper wishes
- Wire can be inserted into the wax to increase its strength
- Gives the bees a head start in drawing out the wax—this allows more time and energy to be spent producing honey, which means more honey for the beekeeper
- The wax is stamped to the size of worker bees not drone bees so the hive will raise only worker bees on the foundation wax
- Makes it easy for the beekeeper to transfer frames between hives
- The foundation in wired frames has superior strength so it is good for when hives are transported around the country for pollination or to follow particular honey flows
- The foundation can be reused over many years
- The strength of the wired foundation allows the frames to be spun at high speeds in a honey extractor.

Reasons for **not** using any sort of foundation in a hive:
- Allows bees to build natural comb to their own unique dimensions, which they have been doing for millions of years in the wild
- Bees use the comb for communication through scent and vibration and naturally drawn comb does this best
- As honeycomb is harvested this allows for the natural cycling of older comb out of the hive
- Foundation is made from recycled wax and can have a high level of contamination of chemicals
- Bees don't need foundation to build straight inspectable comb
- Bees prefer to build their own comb in any space or empty frame if given the chance
- The bees have a specialized and highly functioning wax producing gland and it is more natural to allow them to use it as nature intended
- You have to buy foundation from a specialized supplier and there is extra cost and time involved when buying and fitting foundation into frames.

These bees are slowly building out plastic foundation. Notice the uniform size of all the cells.

Pollen

Grains of pollen are like fine flour but each grain is generally yellow in color. Each grain is so small that it can only be seen under a microscope. Bees collect pollen from a multitude of trees, shrubs and annuals. It is a very important source of protein for the growing young bees. The increase of egg laying by the queen leads to a motivation by the field bees to collect pollen. If you observe worker bees bringing pollen into the hive this is a very good indication that your hive has a healthy laying queen.

Bees have pollen sacks on their hind legs which they use as saddle bags to carry the pollen packages back to the hive. Pollen grains literally jump onto the bee as she visits a flower. Pollen grains have a negative charge whilst the bees have a minute positive charge. Bees can carry their own weight in pollen or honey. In comparison, a jumbo jet is unable to achieve this as it can only be loaded with 25 percent of its own weight to enable it to land and take off safely.

The pollen is mixed with honey and stored in the cells. The pollen quickly ferments and turns into "bee bread". This specialized food is fed to the developing larvae as a protein-packed meal. These pollen-fed bee larvae will increase their weight by 1,500 times in less than one week. The hive will store the bee bread just above the brood area in a half-moon shape. The cells containing the bee bread are only ever three-quarters filled. It is normal to have a rainbow of colors in the bee bread. The color of the pollen changes with the plant species it is collected from. Pollen is probably the most important food source in a hive, especially in spring when brood rearing moves up to top gear in preparation for swarming and summer foraging. By weight, pollen contains more protein than beef. Pollen, naturally sourced from a variety of plants, is the best food for developing larvae and young adult bees.

In times when there is no pollen available, the hive bees will cannibalize larvae and eggs as there are insufficient resources to raise them. Over time bee numbers will decrease and the hive will fail to build up until a new pollen source is located.

The Langstroth frame has had a thin strip of foundation comb inserted as a guide for the bees. The bees have started to build their own natural, clean comb down from this. If there is ample food available this frame can be built out within days.

This worker bee is returning to her hive with a full load of pollen to deliver to her sisters.
PHOTO: *Alphapix.*

Royal jelly

There is one aspect of baby rearing where bees take on a very mammalian characteristic. Bees feed the developing young a white liquid milk-like substance during their infancy. This milk-like substance is called royal jelly. It is a protein-rich food made in the glands of worker bees and placed in the bottom of cells just before the egg hatches into a larvae. Royal jelly is produced in the hypo pharyngeal gland on the bee's head. Royal jelly is high is sugar, protein and vitamins and is a complete food. This is all that the developing larva requires for the first three days of its life. After this time, if the egg is destined to be a worker or drone bee, the diet changes to a mixture of pollen and nectar. If this egg is deemed to be the colony's new future queen, the larva is fed a diet exclusively of royal jelly. This simple act allows the queen to grow to her large size.

Honey

If you spend time observing your bees during summer you may notice that many returning bees struggle to fly straight and even sometimes miss the

CHAPTER 4

forms crystals. Granulated honey does not lose its individual taste but takes on a more grainy texture. Gently warming the honey on a sunny windowsill can return the honey to a liquid state again. Granulated honey is not spoiled, rather it can show that the honey is raw and unprocessed.

Nest scent

In order to successfully raise brood, bees depend on heat. Research has found that bees will try and maintain a brood temperature of 93–94°F (34–35°C). Temperature is increased to ripen honey, often up to 104°F (40°C). During the depth of winter when there is no brood the temperature in the middle of the cluster drops to 72–77°F (22–25°C). Bees must produce this heat by vibrating their wings. Their fuel is honey which they have to consume over and above their own dietary needs to produce heat.

The heat mass of air which enclosed the hive consists of warm, dry air infused with hive pheromones and is germ-free. The heat suppresses harmful bacteria and prevents the occurrence of diseases such as nosema.

Unfortunately the modern-framed hive destroys any retention of nest scent because of the man made "bee space" between and above every frame. Every time the hive is opened by the beekeeper the nest scent escapes and the hive temperature drops considerably and with it all the germ-free, disease-inhibiting scent substances. The bees are forced to work harder to raise the temperature back up to an acceptable level and attempt to regain the communication pheromones in order to function again as a super organism.

As a beekeeper you can choose to use a hive design which mitigates against these hive scent losses. A Top Bar hive and many Warré hives have wider top bars which form a complete seal when in place. When you open your Langstroth hive or Flow hive, use a towel or some hessian (burlap) sacking to cover the top of the frames to help to prevent heat loss.

When inspecting your Langstroth or Flow hive, cover the top of each box with a towel or sacking to help conserve heat and hive scent. This action will also help to keep the bees calm rather than flying around the area.
PHOTO: *Alphapix.*

nectar plants in the area of the hive, a high population of forager bees in the colony, and settled and warm weather so the bees can leave the hive to collect the nectar.

When the nectar flow is on in your area you will notice several key changes to your hive. When you are near the hive on a warm, still day you will be able to smell the curing honey. The bees will be very busy at the hive entrance, leaving and arriving in high numbers. If you look closely you may notice that all the bees leaving the hive are flying in one direction, perhaps to a large nectar-producing tree? If you walk near certain flowering trees and plants in your area you may notice a loud hum as thousands of bees go about their chores, collecting nectar. When you open the hive, the bees will appear contented and busy and may seem to not even notice your visit. Their minds are on the important task of collection and maturing of nectar. On a warm evening you may be alarmed to see what seems like all your bees hanging outside the entrance of the hive preparing to swarm. What is actually happening is the colony is busy evaporating excess moisture from the nectar to turn it into honey to cap. It is called "bearding". A large proportion of the bees leave the hive to help reduce the amount of moisture inside the hive. They also fan the entrance with their wings to remove the moisture of the hive.

The best honey to consume is raw, lightly filtered, local honey. Most supermarket bulk honey is pasteurized by heating. This makes a uniform, long shelf-life product, but destroys all the health-giving enzymes, minerals and individual aromas and taste of the honey. It is normal for raw honey to granulate. Granulation occurs because the sugar glucose is unstable and slowly

This bee is busy collecting pollen and nectar from a borage flower.

then visit that one source of nectar until it has finished yielding nectar and only then will the bees move on to the next species. By visiting this one species numerous times, pollination occurs.

Bees make and store honey as food for the colony over the leaner colder months, when due to the cold weather they cannot leave the hive and the plants are not flowering. Bees will use any excess space to store their honey and beekeepers have exploited this fact by creating boxes which can be stacked on top of each other to encourage the bees to store excess honey. A hive in its first year is using all its resources in comb building and brood rearing and a honey harvest should not be expected.

When beekeepers talk about the "flow" their eyes glaze over as they anticipate copious amounts of harvestable honey. The flow is a term used to describe the time in the year when the majority of the local flora is producing nectar and in turn the bees are turning it into honey. Honey is not produced year round. Bees store nectar as honey when they collect more than they can consume. The key components of a honey crop are obviously the presence of

Uncapped nectar with the comb cells.

insect bodies, harvested from thousands of insects running all over your honey with tiny hooked feet and hairy bodies! Gosh, but isn't it yummy!

To change nectar into honey, the bees convert the nectar sugars into honey sugars and excess moisture is evaporated. Nectar can contain up to 80 percent water whilst honey contains between 17–19 percent water. If nectar is harvested before it is turned to honey it will quickly ferment and take on a very bitter fizzy taste due to the high lactic acid content. Honey takes on its unique flavor and aromas from the nectar source that the bees are foraging on. To obtain a varietal honey such as clover or manuka honey, the beekeeper must move the hive into an area where the predominant flowers are from that one source. The honey is then harvested and processed before the bees move on to the next flowering plant.

The reason why bees make the best pollinators for our horticultural industry is because they are very canny in their foraging. When a bee finds a promising nectar source she will perform a waggle dance to inform the other foraging bees where this nectar source is located. Most of the foragers will

Whilst out foraging, this worker bee has collected bright yellow pollen and packed it in her pollen sacs. PHOTO: *Alphapix.*

entrance hole or landing board. These bees are most likely the foragers who have filled their honey stomachs with nectar and are full to the gunwales. Nectar collected whilst out foraging is stored in the bee's honey stomach. This is actually better described as a crop, similar to a bird's.

A worker bee will visit a flower and, using its long tongue, will drink nectar, filling its stomach. When the bee's stomach is full the bee has almost doubled its weight. Returning to the hive the bee will transfer the nectar to a hive bee by regurgitating it into the waiting bee's mouth. This nectar is then transferred to several bees through the hive. Chemicals in the bee's mouth change the complex sugars of the nectar into simple sugar which the bees find easier to digest. This nectar is then deposited into the waxy hexagon-shaped cells. The nectar is still very runny at this stage so the bees fan the liquid with their wings to evaporate excess water. The resulting concentrated liquid becomes honey, which is then capped with a layer of beeswax and stored until required by the hive. This is normally in winter when flowers are not producing nectar. So basically that stuff you have on toast each morning is sicked-up, regurgitated, chewed, spat out, winged, dried excrement stored in a container made from

YOUR APIARY SITE

If you live in the country then you are probably spoilt for choice as to where to site your hives. In an urban setting a little more thought and assessment needs to go into where to place a hive. When you have bees in a hive and you need to move it, it is not a simple case of lifting up the hive and moving it to the other corner of your garden. The bees have set their navigational compass and all the flying bees will return to the original spot and clump together wondering what on earth happened to their home.

Moving a hive to a new position in your garden

The saying goes you can move a hive 3.2 feet (1 meter) or 1.8 miles (3 kilometers), no more, no less. This is because the bees have set their compass and will return to the original hive position. If for some reason you get your hive placement wrong and need to move your hive, there are a few ways to do this.

Firstly if you have a friend who does not mind a hive visit for a week and they live more than 1.8 miles (3 kilometers) away from you, you can move the hive there for a short time. At night when all the foragers have returned home, block up the entrances and move the hive to the new location and reopen the hive. In the morning the bees will reset their compass to this new location. In a week's time repeat this procedure and return the hive to your garden in its new location.

Another method is to again wait until nightfall, block up the entrance and move it to the new position in your garden. Unblock the entrance but place some twiggy branches or dry long grass in front of the entrance. As the bees scramble over these obstacles in the morning they will take an orientation flight and reset their GPS.

A third option is to again wait until nightfall, block up the entrance and move your hive to the new area. Research has found that the bee's memory only lasts for 72 hours so if you keep the bees contained in their hive for three days and then release them, they will take new orientation flights. I only recommend using this procedure if for some reason you can't use either of the first two methods. Ensure that the bees have adequate food stores and perhaps provide a water source inside the hive. To prevent the hive overheating while it is closed up, use a screen-bottom floor.

When a bee is setting their GPS they make orientation flights. If you look carefully around the entrance of your hive you may witness bees facing the hive as they fly in an extending figure of eight pattern. This is an orientation pattern of flight.

There are too many hives in this apiary site, even if it is surrounded by native bush. Crowded hives leads to drifting of bees to other hives, competition for food sources and the spread of disease.

What makes a good site for a hive?

Sun
A hive does best if it is positioned in a sunny area of your garden. Morning sun is the most desirable as it warms the colony in the morning. All day sun is good but if your area provides some protection from the hot afternoon sun, then that is beneficial.

Wind
A position out of the prevailing wind is required. Bees find it hard to return to the hive heavily laden with nectar in a strong wind. Wind can also funnel through entrances or gaps and chill the colony. If the hive position is in an exposed position, try erecting a wind barrier by planting some tall, quick-growing plants such as bamboo, sunflowers, globe or Jerusalem artichokes as a screen.

Trees
Overhead trees can be problematic as they will shade the hives too much and after rain, the constant dripping onto and around the hives may lead to problems with too much dampness and condensation.

Water
During the brood-rearing seasons a hive requires a supply of water. Bees prefer a nearby water source and will quickly fill and fly back to the hive and unload and repeat these trips throughout the day There are forager bees whose role it is to source and collect water. Bees collect water to dissolve crystallized honey, to water down honey to feed to the larvae and also as an air-conditioning tool to cool down the hive on a hot day. A full-sized hive will require around 35 fl oz (1 liter) of water each day in the height of summer. Bees prefer water from natural sources such as puddles and ponds to clean water. This may be due to the additional minerals they find in this "dirty" water. In saying that, if your neighbor's pool is the closest they will use it. To prevent your bees visiting the neighbors pool install your own bog garden or pond feature or invest in an automated pet watering device that refills itself when the container falls below a certain level.

Water is not stored but used as it is brought into the hive. Once bees find a suitable water source they will continue to visit it. To prevent this, provide a water source in front of the hive. In a large bowl place some pebbles or pieces of floating sticks and keep it topped up. Allow algae to grow and add some compost or even a piece of seaweed. I add a dash of apple cider vinegar and some sea minerals stock supplement solution to this water. If you are going away for a length of time over summer, place this bowl under an outside tap and allow it to drip very slowly to keep it topped up while you are away.

Flight path
Bees come and go from the entrance of the hive. They fly out and up and within 9.8 feet (3 meters) from leaving the hive they have dispersed in all directions up into the sky. This space is known as their flight path. Bees can become defensive if you come too close to the front of their doorway. To prevent altercations with your bees and passersby, position your hive to encourage your bees up into the air as quickly as possible. Do not face your hive next to a footpath or shared driveway or direct your bees into your neighbor's property. There are some simple tricks to encourage your bees to fly into the sky quickly. You can position the hive about 9.8 feet (3 meters) back from a tall hedge to force them up into the sky. You could build a screen using posts and wind cloth to create a barrier.

If your bees are detouring from your prescribed flight path, try turning your hive slightly or placing some tall plants in pots along the side of the hive to direct them.

Other beehives nearby

Commercial operations will load one area with many hives, often over 20 hives. This makes it easy for the beekeeper to visit as it cuts down travel time and labor costs but it is not good for the bees. Too many bees in one area will cause them to become confused and drift to the wrong hive. Competition for the surrounding food resources will be high. Do not keep bees if there are large commercial operations nearby as your bees will suffer. The carry capacity of an urban area is about one beehive to every acre of land. The country landscape can carry much less due to large swaths of grass and monocrops.

Bee number twos

Yes I know you may be surprised but bees do poo and they can make quite a mess. If you have ever wondered what those small, round yellow sticky dots are on your car windscreen, chances are you are looking at bee feces. Bees are incredibly clean insects and would never go to the toilet inside the hive. They will leave the hive to do number twos, usually doing this soon after they exit the hive.

If there happens to be a parked car or your neighbor's newly washed white sheets under them, they don't notice. To avoid this, position the hive so the flight path is not directly over your neighbor's washing line or parked cars.

If you are a clean freak you may just have to accept that as a beekeeper you will have spotty windows and may just have to clean them more often. I think this is a small price to pay for the free pollination and fascinating hobby.

Neighbors

You can't choose your neighbors but it certainly is worth trying everything to get along with them. Your neighbors may not even be aware of hives over the boundary fence. It is only when the bees start causing a nuisance that trouble may begin.

Many people are very fearful of any sort of flying insect, especially the sort which have a sting in their tail. Gardeners normally understand the importance of honey bees in the local ecosystem but unfortunately many people are so removed from the natural world they believe that a hive is a real threat to their family's safety.

Offering free honey can help, as can gentle education about how honey bees operate so as to allay their concerns. Read the situation with respect and

One of my Top Bar hives on our urban property. With careful positioning an urban hive should not cause any concern to neighbors or passing pedestrians.

try to work through any concerns, as good neighborly relations are important in any community.

Working the bees in your garden

When you have your head in your hive, busily assessing how your bees are progressing, you may be unaware if an innocent person nearby becomes an easy target. It always pay to keep gentle bees in any urban area. Monitor your bees' temperament every time you open your hive and if, for any reason, your bees suddenly become very aggressive and it appears ongoing, re-queen at once. The temperament is led by the queen. Within two weeks of re-queening, your hive's temperament should recalibrate and settle. I have had this issue once. One of my Top Bar hives suddenly went from being placid to dive bombing me and everyone brave enough to enter their domain, which suddenly extended to the whole garden. This made gardening and even hanging out the washing very problematic. Of course it was nice to have an excuse not to do the laundry but this situation was hardly sustainable! After removing the queen and introducing a new, mated queen in a queen cage, within two weeks the hive was back to its normal passive, friendly self.

How defensive your bees are can be influenced by the weather and a few other conditions. The best conditions to open your hive are when:

It is important to choose an appropriate time to check your urban hives to avoid any confrontation between your bees and neighbors. PHOTO: *Alphapix.*

- Most of the field bees are out foraging. Normally this is after 10am on a warm day.
- There is a nectar flow on as the bees are busy collecting and there is plenty to go around so there is no threat of other hive bees trying to rob the hive.
- The neighbors are not having a party or mowing the lawn. Noise and vibrations can make the bees more defensive.
- There hasn't been a recent thunderstorm or fireworks as this can stress the bees.
- It is still and warm enough for you to be comfortable in a t-shirt. This is a good gauge for the appropriate temperature to open the hive. If it is too cold the hive bees will be more defensive and the brood can be chilled.
- The bees are not under attack from wasps or other hive robber bees. If a hive is under a constant attack it is naturally more defensive.
- Don't bump or bang the hive as this will alarm the bees.
- Bees react in a negative way to human sweat. Take off watchstraps and have clean clothing on when you check your hive.
- Don't wear perfume or aftershave. Don't eat bananas prior to opening your hive as they give off the same scent as the bees' alarm pheromone.

CHAPTER 5

EQUIPMENT

As with any hobby, you can spend as much money as you wish. In fact all the bee catalogs and supply stores make it very easy to do this. It's not hard to get caught up in the "I need that" trap. You don't have to do this and it is possible to be very frugal when you start your beekeeping journey. Later you can invest in particular equipment if you like it or feel you can't do without it! There are a few pieces of equipment I do recommend and if you are new to beekeeping, these items will give you the confidence to enjoy the bees.

Bee suit

You may have experience with bees in the past or you may be completely new to beekeeping. When I started keeping bees nine years ago I was a total greenhorn! I had no experience and was actually scared of bees. I knew they really hurt when they stung you and thought they were out to get me most times of the day.

To give myself confidence I invested in the uber-protection suit! A full suit with a zipped hood with elasticized cuffs on the wrists and ankles. These suits cost a lot and are heavy to wear and hot to be caged in, but they will give you the confidence you need. Even in this suit I would initially wear jeans, gumboots and a long-sleeved shirt. There is a jacket version, which is cheaper and cooler to wear.

A suit will last for a very long time. Choose one that is baggy as the loose fabric will give your skin protection from stings. It is still possible to be stung through a bee suit but the pain is very much reduced as the sting does not penetrate as far.

There are now suits available that are very cool to wear. These breeze suits are made of three layers of specialized fabric. These suits are more costly to purchase but may be an option if you are a person who really feels the heat.

If you do get stung at any time whilst you are with your bees, smoke the area with your smoker. This will mask the alarm pheromone the stinging bee will emit. This prevents a mob of angry bees charging in to the same site to administer some of their own medicine.

While you are in the hive it is perfectly normal for bees to land on you, often on your hands, arms or back. This is not threatening behavior, the bees are probably just having a rest. Of course if you are new to beekeeping this can be a little daunting, having bees running over your arms. Remain calm

These beekeepers are well-protected with full suits, gloves and their pants tucked into gumboots or boots.

and remember the worst thing you can do is start waving your arms around or start squashing these bees.

Don't store your bee suit in your wardrobe, store it out of the house in a shed or garage. The reason for this is your suit may contain minute amounts of bee venom. These allergens can cause allergic reactions in other family members after constant contact, however minute. For this same reason, when you come to wash your suit do not throw it in with the family wash, rather, wash it separately. This can have the same effect and cause family members to develop life-threatening allergies to bees. The risk is small but it pays to play it safe.

Gloves

At the bare minimum you can use washing up PVC gloves but these do become very hot to wear in the heat of summer. Sweaty hands make for slippery frames.

Beekeeping gloves are a good investment. The better ones are made out of calfskin with thick cotton gauntlets that reach up your wrists and over your forearms before being secured by elastic. The leather makes a good defense from stings. The tighter your gloves the more dexterity you will have but this

These leather beekeeping gloves extend to below the elbow for extra protection. Choose leather gloves if you are able. Over time they will become discolored with the propolis from the hive. Wash them occasionally in warm soapy water with a dash of bleach added to kill any bacteria.

also increases the risk of stings penetrating through the material.

Over time your gloves will get covered in sticky propolis and honey. I normally just throw them in the wash with my bee suit on a warm wash. Periodically it is good practice to get into the habit of scrubbing them with a 10 percent bleach solution to keep them free of any bee disease spores.

With time, as your confidence increases, you may choose to go gloveless. I now always do this, but it did take me over five years to gain the confidence, so don't put pressure on yourself. Without gloves I find I have much more dexterity. I am less prone to squash bees, which leads to no angry bees. If you work slowly and gently, the bees hardly know you are there. Bees spend their lives bumping into their sisters so react kindly if you use your fingers to gently nudge them out of the way. I also find it much nicer not to have any hot, sweaty gloves on. If you puff some smoke over your hands before you open the hive this can help prevent any unwanted stings.

The only time I do normally get stung on my hands is when I pick up my hive tool or a top bar without looking and inadvertently squeeze a bee. If stung, you can quickly flick the sting out, puff some smoke on the area and get on with things.

Hive tool
To open a hive you do require a tool which can prize and separate frames and lids off beehives. The bees will stick everything down with propolis making it almost impossible to do this with your hands. You do not need to invest in a

hive tool. For the first years I just used a bread knife. The only problem with these is, with their dark handles, they are easy to lose in tall grass. I am sure I have a whole cutlery set of sharp knifes lurking in the undergrowth around my beehives.

A hive tool normally has a painted handle (mine is in red) which makes it easier to find in tall grass. Most hive tools have a hook at one end, to lever tight frames out of Langstroth hives.

I would recommend just starting with a rigid old bread knife or perhaps an old blunt chisel. You can buy a hive tool later if you wish.

To prevent any disease transmission regularly sterilize your hive tool by placing it inside a hot smoker and puff until you see flames. The fire will help to destroy bacteria and fungi and to melt wax and propolis off the steel.

The only equipment I now regularly use are a smoker, and a veil to protect my face from stings. As your confidence grows you may choose to not wear a full beekeeping suit.

Smoker

When I first started beekeeping I chose to just use a spray bottle of sugared water to calm the bees. Nowadays I use a smoker.

The theory behind a smoker is the smoke helps to mask the alarm pheromones that the guard bees give off to warn the colony of an attack. The smoke is also said to make the colony think there is a forest fire nearby. This makes them want to gorge on honey in preparation for the escape flight to flee the burning hive. After gently smoking a hive you will see the bees with their heads down in the cells gorging on honey.

When using a smoker, the mantra "less is best" is very true. I may have a lit smoker near me but after a small puff of smoke under the floor or into the entrance, that is often all the smoke I use. It is sort of like a door knock just to let my bees know I am there. Too much smoke has been known to make the bees more hostile.

I feel a smoker is my most important tool. Sometimes I do not need to use it but I have it lit and at hand at all times. PHOTO: *Alphapix.*

Hive designs also dictate how much smoke you may need. In Top Bar hives, a large portion of the colony may not even be aware that one end has been opened. As you are not annoying the whole colony at once, you will find fewer bees flying around you and or even aware of your presence. A Langstroth or Flow hive has differently designed frames. When the hive lid is removed, all the nest scent and heat escapes into the atmosphere. All hive bees are instantly alerted to your presence and depending on their mood, they may be more defensive.

A puff of smoke over the top of the frames can help to drive the bees back down into the hive as you are trying to close up the frames. This can help prevent you squashing bees.

A smoker can be incredibly frustrating to light and to keep lit. There are a few tricks and as with many things, preparation is everything. When you are choosing a smoker, spend the extra money and buy the largest one you can afford. Choose a "commercial smoker". The larger the smoker cavity, the easier you will find the smoker to light and keep going. The key thing to remember is you want the flame to be at the base of the smoker and the dry fuel on top of this. A long-nosed gas lighter makes it easy to get the flame down into the bottom of the smoker. To make your smoker last longer, insert an empty can (such as a baked beans can) with air slots cut into the base and sides inside your smoker.

Lighting a smoker
1. Assemble some fuel materials. Ensure that these are only natural materials. If you are using dried long grass, ensure that is has not died because it was sprayed with chemicals. Avoid color printed paper due to the high level of chemicals. Examples of some fuel materials which are appropriate include:
   - Pine cones
   - Hessian sacking cut into wide strips and rolled
   - Natural string coiled into a light ball
   - Egg carton boxes ripped into pieces
   - Untreated coarse wood shavings
   - Dried, mulched branches
   - Dried pine needles
   - Rosemary, thyme or lavender cuttings
   - Dried cow or sheep dung
   - Corrugated cardboard rolled into tubes (ensure it has not been treated with any chemicals)
   - Dried leaves.
2. Empty out any ash from the last time and ensure that the metal grill in the base of the smoker is replaced right side up with the legs on the base of the smoker.
3. If there is a breeze, turn the smoker so the bellows are up-wind. I have learnt this through experience when a piece of burning material blew out of the smoker and landed on the bellow and instantly melted the leather.
4. Place some lightly crumpled newspaper into the base of the smoker and light. Puff gently until you can see flames.
5. Lightly add some dry fuel materials, all the while gently puffing the bellows. When you see flames licking above the fuel material, add another handful and repeat.

It pays to have all your dry material on hand before you light your smoker. Remember you are aiming to create a bed of embers at the bottom of the smoker with lots of dry material in the top.

The smoke produced from your smoker must be cool on your hand so as not to injure your bees.
CREDIT ALL PHOTOS: *Alphapix.*

6. When you see flames above the fuel material, stuff the smoker full with the remaining material. Puff until you are happy with the amount of smoke exiting the nozzle. Place your hand in the smoke to check that the smoke is cool. You do not want hot smoke or embers coming out of the smoker when you puff. If you see this, stuff the smoker with more material so the embers are forced to the base of the smoker.
7. Give a few gentle puffs at the entrance and wait for 3–4 minutes before opening the hive. If you are using bare hands, a few puffs on your hands will help protect you from any stings.
8. When you are finished using your smoker, use a chunk of dried grass or cork to stop up the nozzle to extinguish the embers.
9. If you are traveling in a car and don't like the smell of the smoker (if your smoker decides to re-light itself on the journey home), store the smoker in a lidded bucket.

Veil

I choose to wear a beekeeping veil most of the time. I find my thick curly hair seems to be a magnet for flying bees. When they get caught they immediately burrow and I am left with a demented buzzing noise and an uncomfortable sting somewhere on my scalp.

I use a wide brim hat with a veil. I like the wide brim hat as it offers some sun protection. Ensure any hat you use with a veil has a brim that keeps the veil off your chin, ears and nose. If these parts of your face are in contact with the veil it means that a bee could potentially sting you through the mesh.

I have a beekeeping friend who has protruding ears. He was always complaining about how the bees would sting his ears through the mesh and how much it hurt. His nifty solution is to wear an elastic headband. He wears it to cover his ears under his veil and it works a treat.

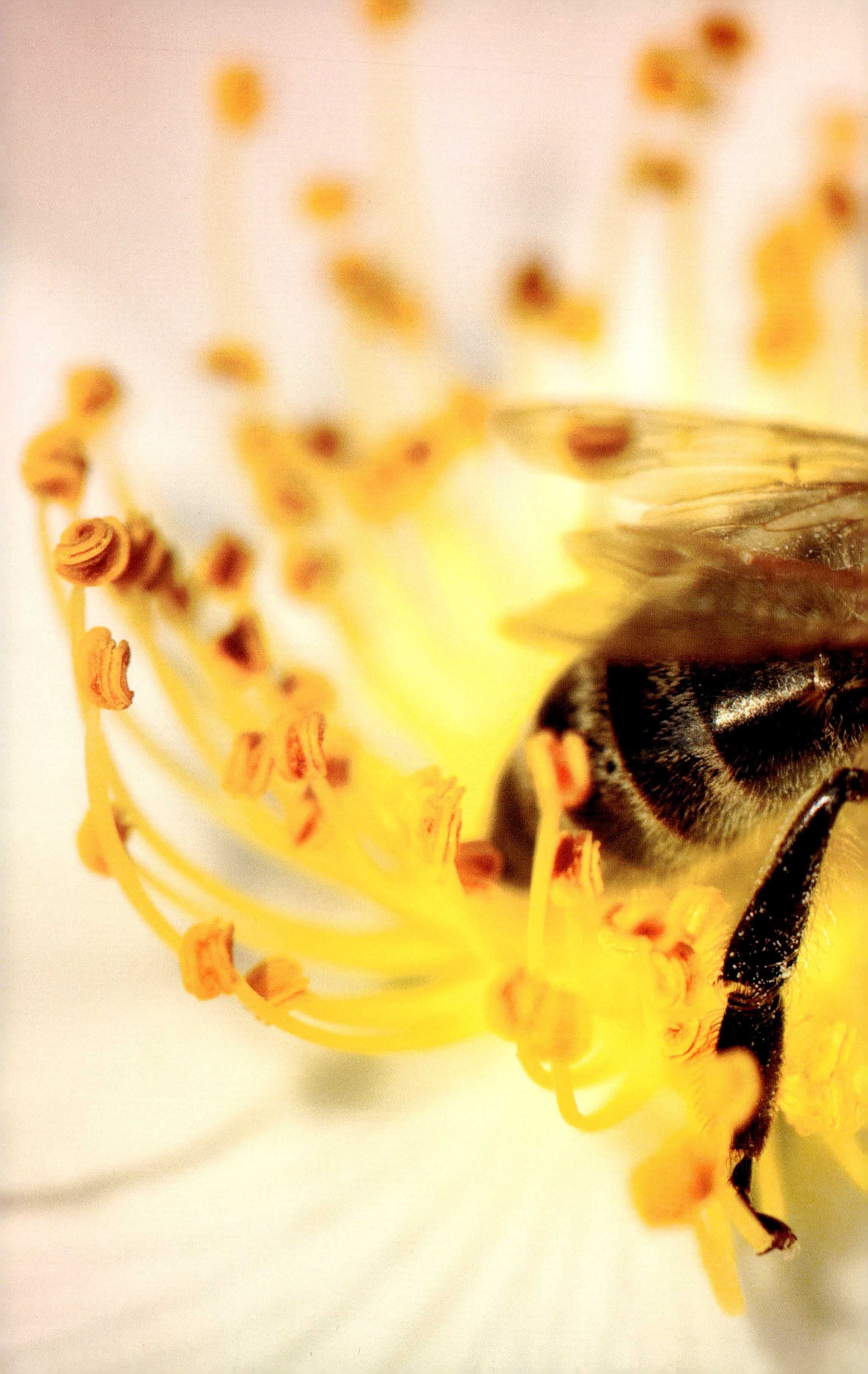

SECTION II

Please remember two things:

1. There is no one right way to keep bees

2. There are no experts except the bees themselves

— Phil Chandler

CHAPTER 6

INTRODUCING OUR THREE NEW BEEKEEPERS

Sarah and her Flow hive

Sarah has always loved honey, and found it fascinating. She became a proper honey convert when she was working in a medical lab and her horse had a lower leg injury, which became badly infected with traditional treatments. After researching wound care treatments she came across recommendations for manuka honey, so she invested in a pot with a high UMF (Unique Manuka Factor) and applied it to her horse's wound. When she replaced the bandage two days later, the proud flesh was gone, the wound had granulated, the manky edges came away clean and the smell was totally gone—it was an amazing transformation.

Before she applied the first manuka dressing, she had taken a swab of the horse's infected leg and the lab she was working in cultured it in their microbiology department, finding all sorts of nasties. The lab tested different strengths of manuka honey on the cultures and found inhibition of growth right down to quite low concentrations—more proof of manuka honey's power. Sarah was hooked: honey was like a liquid miracle for treating wounds and she spent hours reading online articles about honey and how bees use it.

Having always been environmentally conscious, Sarah knew that bee populations worldwide were under stress. She had learnt how important bees are for pollination for human food supplies. So her love of bees and her desire to support and protect them was born. When she met her husband, one of the clinchers was his honey collection. He had at least 110 pounds (50 kg), usually in 1 lb 2 oz (500 g) or 2 lb 4 oz (1 kg) lots, and he'd sourced them from all over New Zealand and had quite an amazing selection of varietals.

Sarah had never kept bees, but she has observed me caring for the Top Bar hive that is kept at the environment center where she is manager. This beehive is in the demonstration garden to help teach people about bees and how they help us. While Sarah and her husband were house hunting for their first home, Sarah knew she wanted a productive garden and to live naturally, and to add to their honey collection with their very own homegrown honey. When Sarah saw the crowdfunding campaign for the Flow hive, a new design of hive which meant no messy, sticky, awkward, stressful honey extraction was needed, she felt this was right for her and her husband Ross. After waiting of many months while the business became operational and started producing the hives, Sarah's hive arrived in early 2016. The main drawcard for Sarah for the Flow hive over other designs was the ease of honey extraction. If this hive hadn't come out when it did, she probably would have been a Top Bar beekeeper.

One of Sarah's main concerns is contracting AFB (American foulbrood). The Flow hive is very expensive and obviously she does not want to see her investment go up in smoke! Sarah is reasonably confident in the ongoing care and maintenance of the hive, and is building up a flow hive network of online support. Sarah planted a small flower garden and some bee-loving plants, which have made the garden more beautiful. Sarah can't wait to show other beekeepers how the Flow hive works. She is really looking forward to her first honey extraction, which will be quite an event!

Sarah lives in a residential suburb on a back section and is surrounded by neighbors. Her property has a large-sized backyard and high fences most of the way around so she is not too worried about the bees bothering the neighbors. Like any responsible neighbor she plans to offer peacekeeping honey. In terms of bee food, there'll be a huge variety within easy flying distance, and the climate is quite mild so Sarah's expectations are that her bees will do really well and prosper.

Sarah has very limited equipment so far, just a veil and a bee suit on order, as well as a hive tool, brush and smoker coming. She is planning to buy anything else as needed.

Sarah has been reading quite a lot about keeping bees, and watching a lot of videos by Flow Hive and on YouTube. She has read some beekeeping books from her local library and will invest in some for the home bookcase. Sarah also plans to join a local beekeeping club.

Sarah has have been the recipient of some quite strong opinions against the Flow hive in some beekeeping circles, which came as a surprise to her. But she is aware of the saying "if you ask four beekeepers a question, you'll get eight different answers". In the meantime she is happy to just ignore the people without hands-on experience and see for herself how it goes.

What is a Flow hive?

To many, the Flow hive looks just like a cuter version of the common Langstroth hive—those stacked boxes which are so familiar out in the countryside. The Flow hive functions like a Langstroth hive. The bottom box contains the queen, comb within frames in which she can lay and the colony can raise brood. The entrance to the hive is at the bottom front of this lower box. As the colony expands during the spring and summer, an extra brood box can be added on top of the lower one to give the bees

The Flow hive.

more room to create brood raising areas and pollen and honey storage. This is an easy way to help prevent the potential for swarming and allows the colony to build up to a strong force to exploit a large nectar flow. When the nectar flow arrives and all the local area plants are producing copious amounts of flowers and nectar to attract all those pollinators, a honey box can finally be added to the top of the hive to give a ready-made pantry for the bees to make and store their honey in. Contained within the honey box are flow frames.

Honey harvesting with the Flow hive
The divergence of this new design comes when the time arrives to harvest honey. Normally the excitement of honey harvesting is tempered with the hard work and mess associated with getting the honey out of the hive into jars. Bees have to be coerced off the heavy, sticky honey frames. These frames are then taken inside and the thin cap of wax scrapped off the top of the honey before the frames are placed in a centrifugal spinner, which spins the honey out of the cells so it can be funneled into jars. The honey is great but the hard work and clean up can be daunting. Expensive equipment is required and defensive bees guarding their precious honey have to be pacified. For a backyard beekeeper there is the issue of either hiring a honey extractor from the local bee club, which comes with the risk of spreading honey bee diseases, or buying this piece of bulky equipment which is only used once a year. This is where the idea of the Flow hive comes into play. Imagine being able to just turn a key and have beautiful raw honey roll out of your hive into jars without upsetting the bees or you even breaking into a sweat.

The Flow hive frames that store the honey are made from food-grade plastic, as are many frames and foundation these days. The plastic makes up the hexagon shape of the comb. When a key is inserted and turned, these plastic cells are split down the middle breaking the cell and releasing the honey to flow down to a collection channel and then out of the back of the hive into the waiting jar. Once all the honey has drained from the frame, the key can be turned to reset the comb and the bees will get busy repairing the wax comb and refilling with nectar, ripening this and recapping. The harvesting process can be repeated as resources allow. The Flow frames can stay within the hive and can be reused. It all does seem too good to be true.

How the Flow hive started
The Flow hive became an internet sensation almost overnight, after its concept

was profiled on a crowdfunding website. Not since the Reverend Langstroth introduced the world to his innovative frame hive in 1852 has there been such a suggestion of innovative change in the way beekeeping can be conducted, especially for us backyard beekeepers.

The inventors of this new system are a father and son team—Stewart and Cedar Anderson who live in Byron Bay in Australia. Being a family of beekeepers, they have been working on this development for the last ten years. As with any new invention the Flow frame has gone through many adjustments and modifications over this time. As with anything new or innovative like this, the knockers and the detractors have been very vocal. In beekeeping, these types seem to be very common and many of the establishment seem to be averse to anything that is not the norm.

Who is it made for?

The type of person who has embraced this new invention is the new beekeeper: the educated busy professional who is looking for an interesting and different hobby which will give them a respite from their busy stressful lives. These generation Y types are willing to try something new and different. They are technologically savvy and willing to make the commitment in financial terms and invest in this new equipment. There is quite a financial commitment, as these new hives do not come cheap. To acquire the full hive will set you back over US$700 and that is not including bees or any other protective equipment.

People who have invested in the new Flow hive are, like many other people, new to beekeeping. They are educated urbanites who are aware of the declining population of bees worldwide. They want to do something about stemming this alarming statistic by establishing a hive in their backyard. Leading busy stressful lives, they want to reconnect with the natural world and beekeeping allows them to do this. Keeping bees makes you aware of weather patterns, changes in the seasons, species of plants and trees, the negative impacts of chemicals and the accumulative effects of these. You can't help but become an environmentalist by default.

Concerns

The beekeeping establishment do have concerns about Flow hives and their potential impact on the health of the honey bees. They fear that these new beekeepers will not be well informed or supported in the maintenance of a colony of bees. They argue that the Flow hive is all about honey production

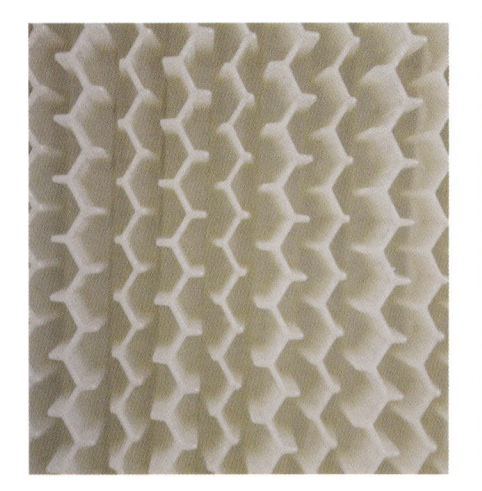

The innovative flow frame showing how the cells open to drain honey.

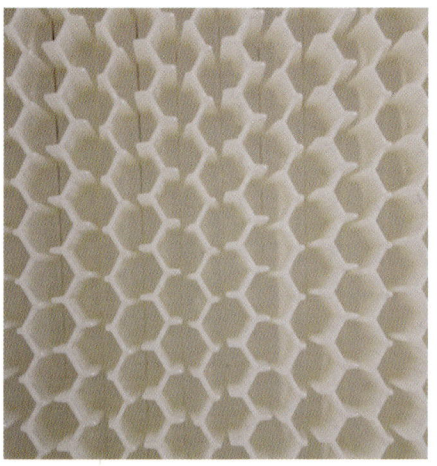

The cells in the closed position.

The bees have waxed up the gaps and will now start using the cells to store honey.
PHOTO: *Alphapix.*

rather than keeping healthy bees. A new beekeeper could keep a flow hive in their backyard and be able to harvest honey without ever thinking they need to open the brood chamber to inspect for diseases and other problems common with hives. Their fears are not ungrounded but it is arrogant to think that these new beekeepers will not take their new hobby seriously. Hopefully this book will go towards helping and supporting just this sort of beekeeper.

I personally believe that the honey bee will be better off globally if there is an increased number of hobby beekeepers keeping one or two hives in their backyards. Surely building a flotilla of healthy, well-maintained hives in backyards in cities and towns across the world is better than having honey bees only located in large commercial apiaries, being worked hard to produce an income for businesses.

The pros and cons of the Flow hive

Pros
- New, trendy hive helps to encourage the tech-savvy, educated and responsible professional to take up the hobby of beekeeping.
- Takes the stress out of honey harvesting, as there is no need for expensive honey harvesting equipment that is only used once a year. This equipment also takes up precious storage space.
- This crowdfunding project has made beekeeping cool and trendy and in the forefront of media.
- The simplified method of honey harvesting may encourage more people to start caring about bees, which will help the insect and have flow-on effects for the environment.
- Harvesting honey via the Flow frames prevents the bees from becoming stressed.
- The Flow frames can be reused year after year.
- Decreased risk of spreading disease, as frames are not transferred to other hives after harvesting.

Cons
- There is a lot of plastic used in the manufacture of the Flow frames. This is bad for the environment and unnatural for the bees.
- Is it just a gimmick as it has had no long-term testing in different countries and with different types of honey?

- It makes it too easy and all the emphasis is on production of honey rather than the time-consuming tasks of cleaning and maintaining a disease-free hive.
- There seems to be no emphasis on managing mites or diseases. Becoming a good beekeeper is not about producing honey but keeping bees alive.
- The hive is expensive to buy. If American foulbrood is discovered in a colony, some countries' regulations state that the infected hive must be destroyed by burning. With such an expensive hive would this encourage some people to just repopulate the hive, thus spreading the infection around the area?
- The "honey on tap" concept could encourage people who are terrified of bees to establish a hive thinking they could avoid regular hive inspections. This could lead to swarming issues and risk of diseases.

Eric and his Warré hive

Eric had always wanted to keep bees, but too many other projects and activities just got in the way. His grandmother was a beekeeper in Holland who grew vegetables and fruit trees as well as raising chickens. Beekeeping was a natural progression after having been a keen gardener for many years.

After practicing as a naturopath in Australia for several years, Eric and his family moved to New Zealand in 1998. They settled in Havelock North in Hawke's Bay on a quarter-acre section, which now boasts many different fruit trees, four vegetable and herb gardens and two Warré bee hives. Eric has planted many herbs the bees love, including lavender, catnip, borage, sage, bee balm, calendula and many medicinal herbs including valerian, liquorice, and echinacea.

All non-organic commercial honey and beeswax will contain, to some degree, a level of chemical mite control contamination, and this is another reason Eric decided to not only keep his own bees, but to abstain entirely from all garden chemical insect and weed control in order to harvest and enjoy 100 percent natural healthy food including honey, herbs, fruits and vegetables. Knowing that his hives are entirely chemical-free he plans to use the wax for candle making and other applications like balms, creams, wood polish and more. Eric avoids all chemicals in his garden and it didn't make sense to medicate his hives yet strive to cultivate quality organic fruits, herbs and vegetables. The other huge bonus in avoiding all chemicals in the garden is that you are less likely to kill your own bees.

Eric's only real concern in his first year was family members or neighbors getting stung!

It turned out that the only person who was stung was Eric, occasionally. Like most beginning beekeepers, Eric had lots of questions and wanted to learn as much as he could and not make too many mistakes.

Eric believes it is important to have a good mentor when you start out with beekeeping; a person who will inspire you and give you plenty of confidence and the ability to get on top of any potential problems before they end up more serious, costing you your colonies if you get too complacent. Don't be afraid to ask, there are plenty of wonderful people in the beekeeping community from all around the world only too willing to give you support and advice.

Eric feels he is most fortunate to learn a lot of very useful beekeeping information from a retired neighbor who used to be president of a local beekeeping association. YouTube is an excellent repository of the most useful beekeeping information and Facebook is one of his favorite resources when it comes to connecting with and learning from beekeepers all around the world.

Eric was drawn to the simplicity of the Warré hive, a hive that is essentially a vertical Top Bar hive design. One of the improvements over the original design includes making frames that are fully enclosed in a cut-down Langstroth frame design. This allows the natural honeycomb the bees make to become more solid and it is less likely to fail with centrifugal honey extraction, because it is fixed to all four sides of the frame instead of just the top bar. Eric recommends cutting-down conventional Langstroth frames as he finds the frames a lot easier to manipulate over Warré's design, which was essentially just a top bar. If you follow Warré's original design, just like a Top Bar hive, you have to be very careful how you manipulate the frames in case the comb breaks off!

Eric enjoys the compact design of the Warré hive boxes and how much easier they are to manipulate and lift even by those who haven't got a lot of strength. A Langstroth super full of honey can be very heavy compared to a Warré box that is full of honey.

Eric has started out with a bee suit, gloves, hive tool and a smoker. Just recently he invested in a centrifugal two-frame manual honey spinner from a trading website. He anticipates it will be a wise investment over the coming months. Eric believes you don't need an extractor until you get several hives.

Eric's recommendations are to invest in beekeeping equipment according to your level of interest. Don't go out and buy everything you think you need until you have kept bees for a few years. It may end up being "not your thing" after one or two seasons and then all your gear ends up on a trading website for whatever somebody is willing to bid for it. (You will often find good beekeeping equipment on websites like eBay.)

The Warré hive

With a Langstroth hive, empty boxes are placed on top of the established box however, with a Warré hive new boxes are always placed underneath or at the bottom of the hive. This is known as nadiring. This allows for the colony to follow their natural instinct of always building downward. Eventually the top box is removed from the hive, full of honey in the autumn.

This simple method allows for a continual cycle of new comb as the older comb is cycled out of the hive without the destruction of the precious brood chamber.

Another design feature of the Warré hive is a quilt box. This is a wooden sided box that has a hessian (burlap) or sacking base and is filled with untreated course sawdust, leaves or straw. The theory behind this is to aid thermoregulation and humidity. As heat rises it is trapped within the natural mass and excess moisture is absorbed.

Pesticides are so prevalent in our environment and the Warré hive removes

Eric gently uses some smoke at the entrance of his Warré hive prior to opening the hive.

The Warré hive showing the quilt box filled with untreated sawdust. This provides insulation and absorbs excess moisture within the hive.

pesticide-laden comb from the hive every few years, making for a healthier hive. The only downside is that all the honey is stored in darkened comb. There is no adverse problem with consuming this comb but if you love comb honey in beautiful white comb, this is not normally possible with a Warré hive.

How the Warré hive started

The design of the Warré hive was the culmination of 50 years of research by Frenchman, Abbe Emile Warré. In the early 1950s the Warré hive was introduced to the beekeeping community and the hive was known as the 'people's hive' due to its ease of maintenance.

Emile Warré studied the life of the honey bee in depth. He analyzed over 300 hive designs ranging from the very ancient pipe and straw hives to the modern frame hives such as the Langstroth hive. He assessed their simplicity of use and ease of management for the beekeeper, and if the particular design aided the honey bee in creating a healthy and natural home. His objective was to design a hive which was easy for a person with only simple tools and carpentry skills to build, was very low maintenance in its care and, perhaps most importantly, mimicked how bees lived in the wild so it would be very bee-focused and natural.

Along with his studies of hive designs, he also spent a long time studying wild bee hives. He discovered that most feral hives preferred to choose a nest cavity of around 12 in (30 cm) squared. He also acknowledged that bees naturally choose to work from the top down. A wild bee colony will attach and build new comb from the top of the cavity. The queen will lay in this comb and over time as more comb is built, the queen will move her laying pattern downwards and the colony will store excess honey above their heads in this older

A Warré hive made from cedar sits on a wooden platform to protect it from ground moisture.

comb. He concluded that the retention of nest scent and heat and allowing the bees to build their own natural comb were some of the most important aspects of natural, low-cost and low-maintenance beekeeping. Empty wooden frames or just top bars are provided for the bees to build their comb on.

While most hive management techniques call for regular individual frame inspections every couple of weeks, Warré believed that it was better to manipulate the hive box by box and only a couple of times a year. In this way the important nest scent and heat was retained.

The pros and cons of the Warré hive

Pros
- A great hive for someone looking to build it themselves
- Low cost as it does not require foundation
- A very low maintenance hive so a good choice if your apiary site is a long distance from your house and you can only visit occasionally
- The design of the hive considers what the bees require to build a hive nest that is as natural as possible
- The continual harvest of older comb is good for chemical-free and natural beekeeping
- Promotes the building of natural comb by the honey bees
- The queen is allowed to travel throughout the hive unrestricted
- Has added insulation qualities so it's good for colder climates
- The ventilated gabled roof helps to shed rainfall quickly and away from the hive.

Cons
- May need to build the hive yourself as this hive design is not readily available from beekeeping stores
- The concept of nadiring requires lifting of boxes so that an empty box can be placed at the bottom of the hive
- Slightly lower honey production as bees use time and energy building natural comb
- In an urban situation the hive can become very high over summer due to its tall upright profile, with a risk of falling over in high winds
- As this is not a common hive design there are fewer experts you can access for mentorship and advice

- As each box is small, at only 12 in (30 cm) square, in an urban situation where nectar sources are plentiful, numerous boxes may need to be added regularly to keep ahead of the space requirements of the bees. This can lead to a very tall hive or a risk of repeated swarming.

A Warré hive looks similar to the common Langstroth hive but the boxes and frames are much smaller and thus lighter to handle.

Sonya and her Top Bar hive

Sonya, her husband Chris and their young family moved back to New Zealand after several years of living a busy expat life in Singapore and Switzerland. Her major desire was to have a garden and keep bees and chickens. She had no previous experience of keeping bees but remembered growing up as a young girl and being near Messop Honey, a honey-producing company in Tauranga in the North Island of New Zealand.

She decided on the Top Bar hive design as she felt it was a simple, natural way to keep bees. She liked how the bees build their own wax and how she could lift individual frames rather than heavy boxes. Her hive has a viewing window and this allows her to view the bees' progress as the colony expands across the hive, without having to open the hive. The viewing window also lets her children and visitors view the bees safely.

With a young family and many visitors, she was concerned about safety so Sonya has opted to keep her bees on a second floor balcony which faces north. This has turned out to be a great location as the bees can fly unheeded from their raised position but Sonya can still watch them come and go. The hive is easy to get to, is safe and can't be knocked over. The location is in the middle of a suburban area so her bees have ample nectar and pollen sources.

Sonya's main concern with her bees is if they will have enough water during the hot summer. She has filled a small pond with water and lays out containers, which get regularly filled by sprinklers in the hot summer.

She loves having the bees in her garden and enjoys seeing and hearing them buzz around the flowering plants. It has changed her thinking about her choice of plants and she now selects plants that are nectar-rich for the bees. She also allows weeds and wildflowers to flower, as she knows how good these species are for providing her bees with nectar and pollen sources.

Before starting out, Sonya read some beekeeping books but she finds it much easier to learn by seeing and doing rather than reading.

She is really looking forward to tasting her first honey but sees this as an extra bonus. Just having the bees in her garden is her main desire.

When she is about to open her hive her main concerns are seeing lots of cockroaches scuttling under the hive lid but she says she is getting used to these. She is terrified of breaking comb so goes very slowly through her hive. She has not been stung but does wear a veil and gloves. She has just bought a smoker but has yet to use it. The hardest thing she finds is returning the top bars to the hive and trying not to squash the bees who poke up their heads between the bars.

> She is still having trouble telling the difference between drones and the queen. Her queen is not marked and she wants to keep her that way.
>
> In regards to treating for varroa, Sonya would like to try the organic options first but is happy to use conventional chemicals rather than risk losing her precious bees.
>
> Eventually she would like to get another hive and this will probably be another Top Bar hive. She has room up on the balcony and says this is working out to be the best spot.

The horizontal Top Bar hive

A Top Bar bee hive is a wooden rectangular box with wooden bars across the top which the bees build comb down from. It is a more natural way of keeping bees as the bees are allowed to build comb to their natural dimensions. The honey is harvested as honeycomb, one bar at a time. The overall design emulates how honey bees live in the wild, using a fallen hollow tree trunk as a cavity.

Top Bar hives have been around for centuries but have been modified over the decades. They are a popular choice in the African continent as they are very easy to build and do not require expensive beekeeping equipment. The hive body can be built out of any scrap timber, no foundation is required and the top bars are again simple slats of wood which sit across the top of the hive body. Over the past decade this hive design has become very popular with backyard beekeepers who want to keep bees in a very low cost, natural way. Proponents of this hive design include Phil Chandler in the United Kingdom and Les Cowder in the United States. Christy Hemenway is another beekeeper who is also a great advocate for Top Bar beekeeping. She keeps bees in these hives in Maine, USA so has experience in overwintering them in very cold climates.

Who is it made for?

If you have been put off by the high cost of getting established with bees this could be the ultimate low-cost hive design for you. I make all my own Top Bar hives and I am certainly no skilled woodworker. I use inexpensive untreated boxing timber for the hive bodies and source untreated wooden pallets to cut down for the top bars.

The other thing I love about these sorts of hives is the way that the hive

A Top Bar honey comb being filled by the bees.

is elevated at waist height. This means no bending or heavy lifting of honey boxes. This is particularly good if you have a bad back or are elderly. I have met many elderly beekeepers who have had to give up their Langstroth hives but are happy to have discovered Top Bar hives to continue their hobby into their retirement. I like to think I am futureproofing my hobby well into my old age too!

This type of hive can suit urban beekeepers as there is no need for extra storage of honey boxes or equipment. This can be an important factor if you plan to keep bees on your apartment balcony or roof space and do not have lots of storage space.

(I have to come clean I am probably slightly biased when it comes to Top Bar hives. I have kept bees in Top Bar hives for about nine years. I have tried other hive designs but this ancient design best suits me and how I want to keep bees. I now have over 15 Top Bar hives dotted around my garden and friends' urban gardens.)

The pros and cons of the Top Bar hive

Pros
- Save money and build the hive yourself. (I make mine out of pallets or boxing timber.)
- No heavy lifting required. This is a big plus for people with bad backs or the elderly.
- No need for expensive honey extractors or other specialized beekeeping gear.
- Considered more natural as it emulates how a hive colony would live in the wild in a hollow fallen tree trunk.
- The bees are generally calmer when you open the hive as the whole brood area is not disturbed. This is good if you have close neighbors or enjoy beekeeping without hot, cumbersome protective clothing.
- Able to observe the bees working without disturbing them.
- No need for wax foundation. As the bees build their own comb from scratch you do not use foundation. The chemicals used in conventional beekeeping for varroa are absorbed in the wax and build up over time. Modern foundation can be full of residual toxic chemicals.
- No need for extra storage for frames and boxes. This can be a real bonus for urban beekeepers with limited room.
- It forces you to be a gentle beekeeper as combs cannot be thrown around or leant against hives.
- Made from thicker wood so are naturally warmer and more insulated than conventional hive bodies. A warmer home is better for the bees, particularly in winter.
- A viewing window can easily be incorporated which is a great way to get kids involved.
- More beeswax is harvested which is ideal if you enjoy using beeswax in craft projects.

Preparing to open a Top Bar hive, but maybe shoes are a good idea?

Checking one of my garden Top Bar hives. PHOTO: *Alphapix.*

Top Bar hive in the garden. PHOTO: *Alphapix*

Honey comb is a lot heavier than brood comb and needs to be managed carefully in a Top Bar hive.

Cons

- The delicate comb must be handled gently. If bumped or held horizontally it will break off the bar.
- The hives are harder to move as they do not stack like box hives do.
- Lower honey production as bees use time and energy building natural comb
- Cross combing can be an issue when a new hive is populated with bees. Careful monitoring is required initially to ensure that the bees are building straight comb so that each comb can be removed and inspected as per some legal requirements.
- There may be fewer experienced Top Bar beekeepers to ask for advice and mentorship. Luckily the number is growing quickly.
- As these hives are generally home-built there is little conformity with measurements between hives. This can make it problematic to transfer combs between hives (This can sometimes be a good thing as it prevents transfer of disease).

This naturally drawn comb of Top Bar honey is ready to be harvested.

Children are often fascinated by bees. A Top Bar hive with a observation window allows them to observe the bees safely.

When you are new to Top Bar beekeeping you will break combs. They are easily fixed by using hair clamshell clips and cable ties. In a very short time the bees will repair and build reattachment comb and you can cut out the hair clip if you wish.

Getting a new colony of bees to build straight comb in a Top Bar hive is important. Placing a thin starter strip of foundation along the top bar and ensuring your hive is level are two successful ways.

CHAPTER 7

GETTING BEES IN YOUR HIVE

Eric and his Warré Hive

Eric has set up his hives in his urban garden. The hives have been placed on solid and level wooden platforms and are sitting in the garden overlooking the sunken lawn. They are facing east so the entrances will receive the morning sun. Behind the hives is a tall iron fence which forms the boundary of his property to the main road, and a pedestrian footpath. As the hives are facing the opposite way the bees exit the hive and gain altitude before flying over the road if they choose to go that way.

Eric took possession of his two newly built Warré hives and had positioned them and levelled them in his garden. A few weeks later Erics was lucky enough to be called about a large local swarm which had landed and wrapped itself around the trunk of a tree. Almost like perfect timing! With the help of a ladder Eric was successful in retrieving this swarm and housing it in his new Warré hive. To move the swarm from the tree into the box Eric used his gloved hands and bee brush to sweep the mass of bees into a large cardboard box. He did encounter a few stings into his gloves which may have meant the swarm had been in that location for over a day and had become cold, hungry and stressed.

When most of the bees were in the box he carefully climbed down the ladder and placed the box of bees on the ground. The bees in the box then started to fan pheromones from their Nasonov gland. This scent alerts all the flying bees that the queen is in this dark box and they will enter the box. Eric chose to wait until nightfall before removing the swarm of bees to his house. This allows all the flying bees to return and settle thus avoiding any small lost clusters of bees still hanging where the original swarm was.

This average-sized swarm is hanging in a lemon tree in the common tear-shaped cluster. The queen will be in the middle of this swarm.

Eric's bee swarm wrapped around a tree. Always remember your safety is more important than retrieving a swarm high in a tree.

Eric settling his new bees into his Warré hive.

Catching a swarm is a good way to start beekeeping, especially if you are starting out in one of the less common hives such as a Top Bar hive or a Warré hive, as these nucleus hives are not commonly for sale. You generally get strong healthy bees, and they are free! A hive will not swarm unless it is strong and robust.

There is a small risk with any caught swarm that they could be infected with American foulbrood. If this is discovered in your hive some countries' regulations (including New Zealand's) state that you need to kill the bees and burn all hive ware. I have caught over 50 swarms over my years of beekeeping and have yet to encounter this risk. To prevent any risk of transferring American foulbrood into a hive from a caught swarm do not insert any drawn out frames of comb into the hive. The reasoning behind this is if you give no resources to the swarm of bees, they need to build their own comb from scratch. Building comb takes a lot of food resources to turn into energy. The honey stored in the bees stomach is used as energy before there are cells to store this potentially infected honey in. Alternatively, hive all swarms in old hive ware that you are happy to part with should it become infected with American foulbrood spores.

What is a swarm?

Swarming is the natural response of a bee colony in order to reproduce. Swarming can mean different things to different people. To the general public it can be a thing to fear and they rush to call the emergency services to get it removed at once. To a commercial beekeeper it can mean the loss of a honey crop as you watch a large proportion of your worker bee force disappear over the trees. To a backyard urban beekeeper is can mean a reason for your neighbors to complain again about your bees or the potential to start a new colony of bees. To many of us it is a wonderful natural sight of Mother Nature in action.

Bees are at their most docile when they are swarming. Before they leave the parent hive they engorge themselves on honey to give them energy during this time. As they are not defending honey stores or brood they are also not defensive.

The preparation and planning for swarming starts around six weeks before the swarm leaves the hive. As the spring weather warms and the forager bees start bringing in pollen and nectar, the queen starts laying drone brood. When you see capped drone brood this is a clear signal that resources are plentiful and the hive may be getting in the mood to start raising more queens. Next, the hive starts building queen cups. These long, peanut-shaped cells are

These are swarm cells which have emerged. Swarm cells are generally found on the bottom edges in the brood area of the hive during mid-to-late spring. If you find this situation in your hive a swarm has most likely already left your hive.

generally found at the bottom of the comb. The queen will select cups to lay a fertilized egg into. The egg-laying queen may revisit these developing queens and halt the development progress by tearing down the cells. Workers do not force the queen to lay eggs into these queen cups nor do they prevent her from destroying their progress if she chooses. The worker bees feed the developing queen larvae a diet exclusively of royal jelly. It is the downward nature of these specialized queen cells which stimulates the worker bees to feed the growing larvae only royal jelly.

The queen begins to lose weight as the workers reduce her feeding. She needs to lose weight to get into flight condition. A queen will lose around one-third of her weight just prior to swarming. Worker bees are flight fit as they are foraging all day but the queen has only ever had a couple of short flights during her entire life—when she left the hive as a virgin queen to mate. Most of her life has been spent being hand-fed at will by nurse bees, and moving slowly over the comb dragging her egg-laden abdomen, laying eggs.

While the queen is on a diet the worker bees are gaining weight. They will gorge themselves with honey to ensure a food reserve for the swarm whilst it is in transit.

Prior to a swarm leaving the hive the colony can be seen having a few practice runs. In the warmth of an afternoon the bees may leave the entrance of the hive and fly around like a dark cloud before returning to the hive after a few minutes. Once the new queen cells are capped the swarm prepares to leave. On a clear, warm, still day the swarm will depart the hive normally in the early afternoon. Bees of all ages will join the swarm but the majority are younger at about four to twenty-three days old. Bees this age are crucial as they have the skills to build new comb. Anywhere from half to two-thirds of the parent hive's population will leave. The swarm leaves just as the swarm cells are capped so if you open your hive and notice capped swarm cells and slightly less mature bees, then chances are your hive has just swarmed.

The swarm leaves the hive and flies en masse in a circular movement before settling somewhere close by to the original hive. Often a tree branch or fence post makes a good resting place. The bees form a tear-shaped cluster, similar to the shape of the rugby ball. The queen will be protected in the center of this mass of bees. Older bees will be on the outside and the younger bees are towards the center with the queen. The inside temperature remains at around 95°F (35°C). The cluster will remain in this location from a few hours up to a few days.

The cluster hangs in this position whilst scout bees are out hunting for a new location to set up a new hive. They search for an enclosed cavity such as a hollow tree stump, the wall lining of a house, under a barbecue hood, a rubbish bin or even a post-box. If a suitable place is found then this communication will be distributed through the swarm and the bees will take off to this new home, normally 1.2–1.8 miles (2–3 km) away from the parent hive.

Meanwhile, in a few days the new queens in the original hive will begin to hatch from the queen cells. The first virgin queen will emerge and seek out other queen cells or newly hatched queens. The strongest queen will kill her sisters and rip down any unhatched queen cells, pulling out the larvae and killing them. The telltale sign of this event is queen cells with a long side exit slit in the cell. If the queen has hatched normally there is a neat round exit hole at the very bottom of the cell. It is often left as a hinged lid.

How to collect a swarm

The best time to collect a swarm is when the bees have formed the initial cluster. If they have settled recently they most likely will be very docile as their stomachs are filled with food. If the cluster has been hanging in this one position for more than a day they will have consumed most of their honey, they could be chilled from being in the open overnight or in poor weather and will become more stressed as the search for a new home is drawn out.

When catching any swarm it is best to demonstrate caution and at the very least wear a veil. You may not know the origin of the swarm and the bees could have aggressive traits. The other safety precaution to always follow is never put yourself at risk of falling if the swarm is high and out of reach. Do not risk an unstable ladder.

The method I use to catch a swarm:
1. Assemble your equipment which should include your bee suit or veil, beekeeping gloves, some frames from your hive, a container such as a cardboard box or plastic bin which your hive frames fit snuggly into, cotton bed sheet and a small spray bottle of white vinegar. Depending on the situation you may require a ladder or some pruning shears.
2. Wearing your protective veil, position yourself under the swarm. Place your container under the swarm and shake or brush the mass of bees into the container. Try to do this in two or three strong shakes of the tree branch. If the bees have settled on a bush or small branch you can cut the

branch above the cluster and gently lower the bees into the box.
3. Spread the cotton sheet on the ground under where the swarm was hanging.
4. Place the box on the sheet. Gently place the empty frames from your hive inside the box, being careful not to squash any bees.
5. Fold an edge of the sheet over the top of the box leaving a small 4 in (10 cm) gap for the bees to enter.
6. Spray the area on the original swarm location with some white vinegar. This masks the pheromones of the swarm so that the flying bees do not return to this area but rather fly into the box with the caught swarm
7. Observe the bees closely around the edge of the box. If you can see worker bees with their abdomen in the air and fanning their wings very fast this is a very good indication that the queen is inside the box. These bees are releasing pheromones from their Nasonov gland, which signals to all the flying bees that the queen is present inside the box. If you look very closely you may see a white stripe on the end of their abdomen—this is the Nasonov gland.
8. Once all the flying bees have settled into the box (often this can be at dusk) wrap the remaining sheet around the top of the box and transfer the swarm to a hive.

Transferring a swarm into your hive

I always prefer to transport the box containing the swarm to my empty hive that night and return in the morning to transfer them into the new hive. It is always easier to do this in the light of day rather than in the cold and dark with tired and angry bees. You can place the swarm box directly in front of or under your hive and open up a small corner of the sheet so the bees can start doing orientation flights in the morning before your return. These orientation flights will re-set their compass to this new hive location.

Process to transfer a swarm box of bees into a new hive:
1. Wear a veil and light smoker.
2. Remove some frames from your empty hive to make room for the frames that the bees are clustering on in the swarm box.
3. Gently unfold the sheet covering the top of the swarm box. Give the bees a very small puff of smoke if you wish.
4. Lift up the first outside frame from inside the box and gently transfer it

over and place in the hive. The bees will be clinging to the frame and may even be hanging like chains from the top wooden bar. If you look closely you may even see some new white comb. If you are transferring into a Top Bar hive, place these frames directly in front of the entrance holes. In a Langstroth, Flow hive or Warré hive, place these frames in the center of the hive box.
5. Repeat this process until all the frames from the swarm box are in place in the hive.
6. You will find that there are still many bees on the base and side of the swarm box. Turn the box over above an open gap in the hive and give the base of the box a strong thump with your hand to dislodge the bees into the hive.
7. Close up the hive and allow the bees to settle and start building new comb for the queen to lay in. During this building time the queen will regain her normal weight.

This bee is fanning and releasing pheromones from her Nasonov gland. If you look closely you can see a white area towards the end of her tail. This is her Nasonov gland.

How to stop a swarm absconding from a new hive

One of my favorite things about beekeeping is catching swarms. Every retrieval is different and presents new challenges. Over the years I have lost count of the number of swarms I have been lucky enough to catch. Nine times out of ten these swarms have chosen to stay once placed in an empty hive. Every so often a swarm will choose just to take off again from the hive. Dewey Caron and Lawrence John Connor, in their informative book *Honey Bee Biology and Beekeeping*, explain that a colony has more chance of absconding if there is no brood or developing queen cells. If the swarm caught is very small (the size of a softball or smaller), are not happy with the new nest site, are disturbed by too much smoke, rainwater drips into the hive, or are exposed to fumes or chemicals, the risk of absconding increases.

If you have just caught a swarm to populate your new hive you want this colony to stay and make this hive their home. I like to introduce a frame of brood to the new swarm from one of my other established colonies. To do this I brush all the bees off this frame and check for any signs of disease before inserting it into the middle of the bee cluster. The swarm bees will quickly cover the brood and clean any of the empty cells in preparation for the queen to lay. There is a small risk of American foulbrood spores being introduced with this new swarm.

Once the swarm has been introduced into the new hive, leave them undisturbed for at least ten days to settle.

Newly built hives from high resin wood such as macrocarpa or cedar can have a strong aroma and I have had instances where a swarm has quickly left and I have put this down to the aroma of the wood resin. If you have a new hive awaiting bees, place it in a sunny area with the inside exposed to the sun to completely dry the wood. It can take a month or more for very new wood to season.

To feed or not to feed a swarm

I never feed the swarms I catch with sugar and water. I prefer to let them build up naturally using the natural food resources around them. Swarms are created from strong hives over times when there is a good source of food. The bees within a swarm are at the optimum age for comb building and it is amazing how quickly a swarm will build comb. It is normal for a new colony to build three full naturally drawn combs each week. If it was a particularly large swarm then they will easily double this. In nature, bees are not fed white sugar and water and as a swarm is one of the most timeless and natural behaviors of bees, then I believe it is best not to mess with Mother Nature.

Monitoring your hive as they build comb

In any hive which encourages natural comb it is important to closely assess the new comb building. Because they can build very quickly it is important to monitor the comb closely to ensure that they are building the comb straight on each individual wooden frame. If the comb starts to get a bit wonky and off-center, then it is really important to gently manipulate the comb back into the midline as soon as possible. If the comb is not corrected the bees will continue to build following this new pattern and very quickly you can get a real mess of cross combing.

Eric checking his Warré hive bees. In this frame the bees have built new comb and are starting to store pollen in the right side cells. Warré frames are much smaller than the standard Langstroth frame.

New comb being built by the bees in Eric's Warré hive. Bees will always build from the top down.

Cross combing is when a comb is fixed to several bars making it impossible to remove the frame without destroying the comb. When comb is new and white it is very soft and can normally be gently pushed into shape. If you address this problem quickly then a much bigger mess and the destruction of this valuable comb can be avoided.

Many countries, like New Zealand, have regulations which state that whatever hive design you use you must ensure that each individual frame of comb can be removed and inspected for diseases.

Varroa treatments for your new swarm

Taking possession of a new swarm presents you with a unique opportunity to treat your new bees for varroa and have a very good knockdown rate using one of the softer organic treatments. I treat all my newly established swarms with one oxalic vaporizing treatment before there are any sealed brood present. This is because this particular organic treatment will not penetrate through the wax capping to kill developing varroa in the cell with the growing larvae.

If you do not know the treatment history of the swarm you caught then it is good to take the opportunity to treat for varroa. This will mean that the colony will enjoy good health and an absence of viral illnesses, which varroa spread, as they start the busy time of building comb, brood raising and nectar collecting, all before winter descends.

If you choose to treat your hive with one of the conventional chemical strips, insert this into the brood area between the bars. Make a note of the date and remember to remove the treatments after the set time, generally six weeks.

Getting bees into Sonya's new Top Bar hive

Sonya had her new Top Bar hive set up and ready for bees in early spring. She chose the perfect urban position for a hive. She has positioned it on the second story of her house on a small balcony off a bedroom facing the morning sun. This is a great choice as the bees' flight path is raised well above any passing people. As it is elevated, many visitors and neighbors are not even aware that it is a hive.

Tom Seeley is a professor of neurobiology and behavior at Cornell University. He has an informative book about how honey bees function and in particular how and why they swarm and how they choose a new nest cavity. His book is called *Honeybee Democracy*. Following the advice in this book, Sonya set up her Top Bar hive as a bait hive, hoping to attract a passing swarm to take up residence in her new hive. Basically a bait hive is a particular-sized, enclosed waterproof container set in a location where bees may discover it and direct a swarm to move into it. If the stars align you can get bees for free and have them all set up in a hive.

Sonya placed fifteen empty top bars behind the hive entrance and enclosed this internal space with the two hive follower boards. To make it attractive to bees she rubbed the insides of the hive with some old darkened brood comb and placed three drops of lemongrass essential oil on to the inside of the hive. This lemon smell mimics the queen's pheromone. As her hive is situated about 16 feet (5 meters) from the ground we hoped that it would be a good attractant to early spring swarms. She reapplied the drops of lemongrass oil every week to recharge the aroma.

Building your own swarm traps

If you have bees in your area and want to start your own colony, why not try and lure a swarm? It makes sense to incorporate the frames of your hive into the bait hive so that when/if you catch a swarm you can easily transfer the bees over to your permanent hive, frames and all. Following the research by Dr Seeley into what makes a successful bait hive, follow these key points:

Key design criteria for a bait hive
- Make the internal dimensions between 10.5–15.8 gallons (40–60 liters). This is the size preferred by scout bees, as it will give the colony ample room to expand and store honey for winter. In *Honeybee Democracy*, Thomas Seeley describes how scout bees measure the internal cavity by walking on the walls and flying to the opposite wall to measure the volume. To help you visualize what 10.5–15.8 gallons (40–60 liters) looks like, a ten-frame Langstroth medium super is 8 gallons (30 liters), a Langstroth nuc box is only 6 gallons (23 liters). A ten-frame Langstroth super is 11.4 gallons (43 liters).
- Make a single entrance of 2 in x 2 in (5 cm x 5 cm) towards the bottom floor of the container. This mimics the entrances commonly found in natural hollowed trees. This size is easy for the bees to defend. You can

cover this hole with some chicken wire to stop birds entering and nesting in the cavity.
- Build to last. As you may want to use this bait hive year after year, build it to last. Its success rate will increase with age as it absorbs the scents of past swarms thus making it more attractive to swarms.
- Make the bait hive lightweight but robust. Perhaps use thin plywood. It needs to be light so you can safely climb down a ladder holding it under your arm when it is full of bees.
- Make it weatherproof as it needs to stand up to rain and wind. Ensure that it is totally waterproof as bees will not choose a wet or drafty cavity.
- Make it out of cheap wood or recycled wood from pallets as it may be vandalized or fall from heights. The cheaper it is to make, the more traps you can make and spread around your neighborhood.
- Make it with a removable hive roof so you can easily transfer swarms into your new hive.
- Design it with some wooden handles or similar so it is easy to attach to a tree and remains secure and safe. You don't want a box of angry bees falling on anyone.
- Paint it white or another light color to keep the hive cool.

These three stacked Top Bar hive swarm traps have been built out of recycled pallet timber. Being low cost it matters less if they are stolen or vandalized. Once they have been painted and the top bars are in place, a lightweight waterproof roof will be installed and they will be put out in various locations around this beekeeper's neighborhood.

Where to place your bait hive

To catch a swarm of bees you need a hive of bees somewhere in the near vicinity of your bait hive. If you see bees on flowering plants around your neighborhood then you will have nearby hives somewhere. Following Thomas Seeley's research, these are a few key requirements to increase your capture success:

- Place bait hive at a height of 6–10 feet (2–3 meters) above the ground.
- Place the bait hive in dappled light but never full sunlight as this could cause the hive to overheat. Under a tree canopy is a good location.
- Place the bait hive near an apiary, on the edge of woodlands or near to a water supply.
- Ensure that the bait hive entrance is easy to see and not obstructed by vegetation.
- If you have observed swarms in a certain tree, season after season, then this would be a good location for a bait hive as for some reason this is a good location for swarms.
- Introduce the correct smell to your bait hive. Smear old comb, propolis and lemongrass oil onto the inside of the hive. A cloth soaked in lemongrass oil placed in a ziplock bag with a small slit is a good option if you can't visit the location regularly to re-scent. If you don't have access to any old brood comb buy some beeswax from a candle supply shop.
- Take down the bait hives at the end of summer and store in a covered area until next spring to extend their usefulness.

OMG! I've caught a swarm of bees in my bait hive—what do I do now?

If you see about ten bees buzzing in and out of your bait hive, this may just be scout bees checking it out as a possible new home. If you see a constant stream of bees coming in and out of the bait hive and the activity at the entrance seems constant and busy throughout the day then you probably have been lucky enough to catch yourself some bees. Well done! Now what?

Firstly, leave the bait hive in position for at least 24 hours to ensure that the swarm has settled. If your bait hive is already 3 miles (5 km) away from its permanent position then you can simply move the bait hive at night to its new location. If the bait hive is close to where you want to house them, then follow these suggestions. The hive can be moved to a more permanent position in three ways:

1. Carefully take the bait hive down and settle it on the ground under the original location. Every second day move the bait hive 3.3 feet (1 meter) each day towards the final site. This method is satisfactory if you only need to move the bait hive to the other side of a garden or similar.
2. At night, once all the foraging bees have returned to the bait hive, take down the hive carefully. Remember the hive will be much heavier as it is now full of bees and new comb. Move it to a new location at least 3 miles (5 km) away and leave it there for a week before moving it (again at night) to its permanent position.
3. At night, take down the bait hive and move to the new position and cover the entrance of the hive with branches and leaves. When the bees start flying in the morning they will have to struggle through this obstacle and most will notice a change in the entrance and take reorientation flights and return to this new location. Have an empty box at the old bait hive location to catch any returning bees and empty these into the hive each night. You may need to repeat this over a few nights until these bees learn their new location.

Sonya and her Top Bar hive

Six weeks went by and not a scout bee was to be seen checking out Sonya's hive so, as the season was progressing, she decided to purchase a nucleus of bees to start off her new hive.

A nucleus (nuc, nuke, nook) is simply a mini-colony. A nucleus contains a young, newly mated queen, some drawn-out comb which contains capped and uncapped worker brood, and stores of pollen, nectar and capped honey. A nucleus can be as little as two frames of comb up to the more common five frames. It has the population and resources to start building into a large colony through the season.

When buying a nucleus colony, ensure you buy one which has the same design as the hive it is being transferred into. For instance you can't transfer a full depth Langstroth nuc into a three-quarter depth Langstroth hive as it simply won't fit. Similarly if you buy a Top Bar nuc, as Sonya did, check with the seller that the comb dimensions are the same as your hive as there is no uniformity of design with Top Bar hives.

Questions to ask the seller of a nucleus beehive:
- How old is the queen?
- Is she marked?
- How many capped and uncapped brood frames are included with the nucleus?
- Are the frames newly drawn or older brood frames?
- Are the frames plastic or wooden?
- When was the nucleus colony last checked for American foulbrood disease?
- When was the colony treated for varroa and what with?
- Is the beekeeper registered and have they been trained to recognize American foulbrood disease?
- Is the nucleus box included in the price of the nucleus colony?

A nuc will require feeding with sugar water if there is no nectar flow. Feeding sugar syrup will stimulate brood production but it is best if the nuc has at least one frame of capped honey and some cells of pollen to help it get off to a good start and provide food and insulation in case of cold night time temperature drops.

When Sonya received her new nuc she immediately placed them in her new hive. She did this by creating a space right behind the entrance holes, which on her hive are in the middle of the hive body. She carefully lifted out each individual top bar from the nuc box and placed it in the same orientation and position into the Top Bar hive. The two end top bar frames have to be very carefully moved as when bees have been in a small nuc box they can quickly build some extra bridging comb onto the ends of the nuc box. Sonya cut this bridging comb with a long bread knife.

Careful and considered movement is called for to transfer the fragile comb and bees into their new hive. With all Top Bar hive combs it is important to also keep the comb hanging on the vertical. At any time the comb is turned even slightly horizontally there is a potential for the comb to break from the top bar and collapse. Sonya also spotted the young unmarked queen, which was reassuring for her.

Once all the nuc frames were in the new hive, she placed two empty top bars on either side of the cluster of bees. This allows the colony space to start building new comb in either direction. Sonya can expect that the nuc should build up rapidly as they have food and brood of all ages so there is little interruption to the colony's advance. With a newly mated queen and

Sonya carefully placing her new nuc of Top Bar bees into her new hive on her balcony.

good forage nearby it is not uncommon for the nuc to reach its full production strength at the end of its first year.

Sonya chose to feed her bees to ensure that they had adequate resources to increase their numbers. It is best to always feed bees inside the hive as this prevents wasps, ants and bees from other colonies being attracted to the hive and robbing it of its food resources. To feed in a Top Bar hive, an easy method is to fill a ziplock plastic bag with sugar water, place it on the base of the hive next to the colony and then snip a small ¾ in (2 cm) slit in the bag. The bees land on top of the bag and drink up the nectar, and their weight pushes the bag down as the fluid is used. Another way is to use a plastic container filled with sugar feed, covered with some small twigs or a kitchen sponge. The bees can land on these "rafts" and drink the feed without drowning. Bees will drink a lot of sugar water and it is best to check and refill as necessary every two to five days.

To feed or not to feed?

I prefer not to feed unless it means the bees will starve and die without feeding.

This is the queen in Sonya's nucleus. She has a long abdomen which is normally darker and not striped like the worker bees which are surrounding her.

I feel the best food for bees is honey, not white sugar. Research has shown that constant feeding of white sugar changes the pH inside the hive, which can upset the natural and fragile balance of the colony.

It is very rare that an established hive will require feeding as long as the beekeeper has not taken all their honey reserves from the last season. In the 2015 New Zealand Colony Loss and Survival Survey of hives by Landcare Research, it was found that over 92 percent of large commercial hive operators feed their bees white sugar feed as a supplement (Brown and Newstrom-Lloyd, 2015). This may be because there is so much money in honey. As backyard beekeepers we are in the unique position to practice a more bee-focused type of bee management and leave sufficient stores of the bees' own honey for our winter bees.

Honey, without question, is the best food for bees. It is a much more complex carbohydrate than white sugar. Only feed your bees their own honey or honey you can guarantee is free from spores of American foulbrood. Never feed your bees store-bought honey as this has the potential to spread this disease, resulting in the loss of your precious bees and hive.

Making a 2:1 sugar syrup feed

Method
- In a large pot pour in 35 fl oz (1 liter) of hot water, add 4 lb 8 oz (2 kg) of white sugar (do not use brown or raw sugar as these can cause diarrhea in the bees).

- Heat gently until all the sugar has dissolved. Do not allow to boil.
- I add ½ fl oz (20 ml) per liter of animal stock seaweed supplement so the solution looks like very weak tea. This supplement adds vital trace minerals and vitamins the bees can use. This can help with brood production and increased hive health.
- Store in an airtight container until required.

Sonya placed a plastic container next to the end of the cluster of nuc bees and enclosed the hive with the other follower board. It is best to keep the floor closed if your Top Bar hive has a hinged bottom floor. With such a small colony of bees anything you can do to help them stay warm and maintain the hive temperature of 95°F (35°C) is best. To prevent any chance of robbing from a stronger hive nearby it is also a good idea to reduce the entrance either with corks, if your hive has round holes, or with a strip of wood.

Sonya chose to feed her new nucleus colony some sugar syrup. To prevent bees drowning she used a plastic container and a kitchen sponge. This container required topping up every two to three days. The feed was positioned next to the cluster of bees.

Transporting a nucleus colony
- Wedge the nuc box between two items to keep upright if transporting in a vehicle
- Do not leave in a hot car as the bees can quickly overheat
- If you need to leave the car during your journey home, park in the shade and leave windows open slightly
- Use air-conditioning to keep the colony cool. When the bees are cool they will be quieter and more docile.

Transferring the nucleus colony into a hive
- Transfer from the nuc box as soon as possible.
- If arriving home late, you can place the nuc box next to the hive and open the entrance to allow the bees to orientate themselves to their new location. In the morning you can then transfer the bees into the hive.
- Transfer the frames into the hive in the same orientation and positioning they came out of the nuc box.
- Give at least two empty bars either side for the bees to work on and then close with follower board if your hive design uses one.
- The bees will require a lot of food resources to build new comb. If a nectar flow is on in your area you may choose not to feed them.
- Reduce the entrance of the hive to make it easier for the small colony to defend.
- If you choose to feed them a sugar syrup, I give a 2:1 solution of white sugar (dissolved) to water. Feed inside the hive to prevent robbing from other bees.
- Check regularly (at least every seven to ten days) to ensure that the bees are building straight, easy-to-inspect comb.
- Ensure that your hive is level.

Sarah and her new Flow hive

Prior to receiving her new Flow hive we had hoped to capture and hive a swarm of bees so that when the new hive arrived we could transfer a strong new colony into it. Unfortunately swarms were not common this spring so we had to go to plan B, which is obtaining a split from a fellow local beekeeper, Kevin. I had some queen cells available but Sarah required a nucleus colony that was on the same-sized frames as her Flow hive was designed for. These are full depth Langstroth frames.

Obtaining a split
Kevin visited his hives and chose three frames from within the brood area that would be good candidates to form a split from his hive. He chose frames from one of his strongest colonies. He was looking for frames full of capped worker

Kevin checking a frame of worker brood. This frame is a perfect candidate for a nuc as it is full of sealed worker brood. When this frame of brood hatches it will give the small colony a real boost in population. This frame shows a great laying pattern as there are not too many empty cells.

brood. Making a split in a strong robust hive is a good way to prevent bees from swarming as by taking some of the population of worker bees and food resources it gives the hive extra space and work to build new comb.

To make a successful split it is important to choose frames of a particular kind. The first two frames Kevin chose were frames full of capped worker brood. A frame of capped worker brood does not require feeding from the hive bees and only requires them to keep the developing larvae warm. All the feeding work has been done during the time that the larvae are uncapped and growing. Soon they will hatch and start work immediately inside the hive as hive bees looking after the new queen. A final frame full of pollen and honey was chosen to provide food stores to the young colony until it builds up.

These frames are removed one by one from the parent colony and before they are placed into a small nuc box are checked for the queen and any queen cells. Kevin did not want to leave his large colony queenless. If queen cells

Kevin carefully checks each frame for disease before transferring it into the nucleus hive.

This is an in-hive feeder which is designed to fit inside a full-depth Langstroth hive. It is filled with sugar syrup. The bees will drink from this feeder. To prevent bees drowning, some black plastic netting has been inserted inside the feeder so the bees can climb back up this.

were found these could either be removed and added to further nucleus colonies or left in this nucleus colony thus mitigating the need to introduce another acquired queen cell. The other very important inspection is to check each frame for disease, in particular American foulbrood. It is important for any beekeeper to get in the practice of checking for disease whenever a frame is removed and transferred into another hive.

The three frames were placed into a small nucleus box along with an in-hive feeder which was full of sugar water. Lastly, a frame of foundation was inserted to fill the space as most nucleus boxes are designed for five frames and if you leave an open space the bees can start building crazy comb in this space.

Misshaped comb has been built by the bees at the end of this hive. To prevent this always ensure that the hive is filled with the appropriate number of frames before you close the hive. This comb will need to be scraped away next time the hive is opened in order to remove the frames.

This protected queen cell is being gently inserted between two brood frames in the middle of the brood nest. The hive is then carefully closed and left alone until the queen has hatched.

The nucleus is then moved away from the parent hive so that the flying bees do not return to the original hive and leave the nucleus colony without foraging bees. It is best to move it away at least 1.8 miles (3 km) if at all possible.

Kevin chose to leave this small nucleus colony alone for a few hours to settle and allow it to realize it was now queenless. This only takes a few hours. In this time the queen's pheromones have dropped and the bees are desperate to make a new queen from any appropriately aged larvae. At this stage a bought or acquired queen cell could be inserted onto a frame. To ensure that the queen cell was not damaged by being squeezed between frames, dropped whilst being inserted, or ripped apart by the worker bees, we enclosed the precious cell within a protective cage. You can buy these from beekeeping stores but I like to improvise and use hair rollers.

The protected queen cell is carefully placed on the face of a brood comb approximately 2 ¾ in (7 cm) down from the top of the frame. It is then gently sandwiched with another brood comb. Being placed in the middle of the brood area and down from the top of the frame ensures that the cell remains warm. It is good practice to refrain from opening the hive for the next three weeks, allowing the hive peace to be introduced to the new queen and for her to take her mating flights. If the hive is disturbed the colony has been known to turn on the queen and kill her.

All going to plan, this queen cell will hatch at day 15 or 16. The new queen will stay within the hive for a few days to mature and then on warm, sunny days she will commence her mating flights. She will take these over several days. She will then be a mated fertile queen and commence laying. It can take a few goes for the queen to get the hang of this egg laying thing and it is not uncommon for her to lay multiple eggs in one cell initially. Things should quickly settle down as she gets into her stride.

Placing a nucleus into a Flow or Langstroth brood box

A nucleus will need to be transferred into the larger brood box so that the bees can start drawing out new frames of foundation. The queen can then start laying on more comb so that the colony can expand its population. The brood box sits at the very bottom of the hive below the flow box.

To transfer the nucleus hive you carefully need to shift the frames covered in bees from the nuc box into the brood box. It is important to do this so that the frames are in the same orientation and position. A colony of bees will always build out from the middle of the brood box so place the nuc frames in the center of the brood box with empty frames either side. Fill the box with frames, do not leave any empty space to prevent the bees making crazy comb in this vacant space.

What you need
- Nucleus full of bees
- Brood box full of empty frames
- Smoker, hive tool and protective gear
- Hive floor, hive lid and crown board.

These queen cells have been raised by a professional queen breeder. A queen cell looks like a peanut shell. Handle it very carefully and always hold it upright. The cell must be kept at around 95°F (35°C) prior to being placed into a queenless hive.

These bees are building natural comb on a Langstroth frame. There is a strip of wax foundation which has been affixed to the top bar as a guide. The bees band together forming a chain-like structure to measure the internal space whilst they produce and pass tiny pieces of wax flakes between them. It really is a miracle in the making.

Method
1. Wearing protective gear, give the nuc hive a quick, light puff with the smoker.
2. Place the nuc hive next to the brood box, which is in position and sitting on a hive floor.
3. Remove the middle frames from the brood box to make space for the number of frames in the nuc box
4. Using a hive tool or your fingers, gently lever out one of the frames right next to the side of the nuc box.
5. Carefully place this frame with all the bees clinging to it in the brood box in the same position and orientation as it came out.
6. Repeat this process with all the other frames.
7. Reposition the empty frames so that the nuc is nestled in the middle of the box with frames tight up to it on each side.

If your hive becomes queenless, it is possible to order a mated queen from a queen breeder. Your new queen will arrive in the post in a special cage escorted by worker bees. A candy plug allows the new queen and the queenless hive to get acquainted slowly through the cage as the candy plug is eaten away by the colony.

8. There will be bees inside the nuc box so turn it upside down over the top of the brood box and give the base a good slap with your hand to make these bees fall out and into the cluster of bees in the brood box.
9. Gently brush any bees with a clump of long grass, a small branch or a bee brush so that the sides of the brood box are free of bees then slide the crown board on to the top of the brood box. (If you slide rather than place the crown board you will push the bees out of the way rather than squashing them.)
10. Place the hive roof in place.
11. If there are still bees in the nuc box leave this in front of the hive overnight so these stragglers will go into their new hive.

CHAPTER 8

SEASONAL MANAGEMENT IN YOUR FIRST YEAR OF BEEKEEPING

Spring management in your first year of beekeeping

Spring is all about building up a healthy population of worker bees which can care for all the new eggs and larvae the queen is now producing. With the warming weather, a robust, foraging population of bees gather pollen as a protein food for the growing larvae, and nectar to turn into honey so it can be stored and used as required.

Your new hive is busy building new comb to raise brood in and providing plenty of food. A single open cell larva within a hive is visited 1,300 times every day by nurse bees to feed it.

It is a good idea to inspect your hive every two weeks as a new beekeeper. Often this is more for your benefit than the bees. I think it is important to get familiar and confident with opening your hive and assessing what is going on. Beekeeping is a very visual hobby and it does take time for you to start to recognize what is going on inside your hive. Initially it just looks like a mass of bees crawling around on the frame but soon you will be able to pick out the different types of bees, maybe even find the queen, differentiate between worker and drone brood and even spy newly laid eggs and the older larvae floating in their pool of royal jelly. Too many hobby beekeepers do not spend enough time assessing their hive and leave them unmanaged at the bottom of the garden. This can lead to repeat swarming or the spread of disease. During your first year I really encourage you to get into your hive every two weeks to assess what is going on. When you get more familiar you can afford a more hands-off modus operandi next year. When you understand why you are doing something then often it may not be necessary to do it.

An inspection is done to check on the health of your bees, that the queen is present and laying, that the new colony is building up and has adequate resources, and there are no signs of disease. Keeping a diary to record the date of each inspection and any observations is a good idea. There are also phone apps you can download to do this.

When I talk about a hive inspection often this does not mean going through each and every frame but rather a quick check of a couple of frames until you are satisfied everything is fine.

How to inspect your new hive

1. Choose a day that is calm, warm and sunny. A temperature where you are comfortable in a t-shirt is a good indication that it is warm enough to open a hive. I prefer to do most of my beekeeping around 10am, after most

Over time, with patience and practice your visual assessment skills will allow you to read what is going on in each comb you inspect. PHOTO: *Alphapix.*

of the foragers have left but before the heat of the day when combs can become more fragile.
2. Assemble your protective clothing and beekeeping gear and light your smoker if you are using one.
3. Walk to the front of the hive and stand to one side. Observe the entrance. Can you see bees coming in with pollen loaded on their legs? This is often a good indication that there is a laying queen as the new larvae require the pollen as protein food. Calm, busy activity, with bees coming and going from the entrance is a good sign. Bees coming back looking like they are very heavy, landing and walking into the entrance could mean they are full of nectar. You may see a bee leaving the hive carrying a dead bee. This undertaker bee is doing routine hive cleaning duties. *If* you see many dead bees being discarded from the hive then further investigation is in order as there could have been a mass pesticide poisoning incident. Crawling bees outside the front of the hive could suggest a high level of deformed wing virus, something that is transmitted by varroa mites when their level is high in the hive.

4. If you see bees fighting each other at the entrance and bees darting around then making a dash for the entrance, this could mean robbing from another colony is taking place.
5. If your hive has an observation window, open this and observe what the bees are doing. Can you see new comb being built? It will be bright white. You may see that the cluster is moving across and becoming larger and new bees are hatching.
6. If your hive has a removable bottom board, look at this. If you see tiny flakes of white comb then this is a clear sign that the bees are comb building. You can often see small balls of pollen on the floor. Can you see any varroa mites? If you see flakes of brown wax this could mean that there has been a large hatch of new bees as the wax cappings have been chewed off as the bees emerge.
7. Give the hive a few small puffs of cool smoke at the entrance. Wait two minutes and then gently remove the roof, avoiding any loud bangs or vibrations.
8. Gently inspect the frames. Avoid jerky or rapid movements. Work from the side or back of the hive, out of the way of the entrance.

For a Warré hive and a Flow hive

1. Start the inspection by carefully removing the second frame to the side of the box. The most outer frame is often glued to the side of the box. Be careful not to roll or crush bees. Check this frame and then carefully lean this frame up against the outside of the hive. Leaving this frame out while you are conducting your inspection gives you more room to move additional frames.
2. Gently place the hive tool under the end of the frame and prize upwards so it is possible to grasp the frame. Repeat this at the other end of the frame. It is common to hear a crack as the propolis seal is broken. Hold frames securely over the top of the hive so if any bees, in particular the queen, fall off they will fall back into the hive rather than getting stepped on.
3. You can rotate the frames to view the other side.
4. If the frame is covered in bees and obscuring your view of the cells then give the frame one sharp shake over the hive to dislodge the majority of the bees.
5. It is a good practice to cover the top of the hive with a square of hessian (burlap) sack to contain the hive scent and temperature.

If the hive consists of more than one box, start at the bottom box and move upwards. If the bees become overactive and start bombing your veil they are becoming defensive. It is best to quietly close up the hive and return another day.

Conducting a Top Bar hive inspection

1. Smoke the entrance of the top bar. If you have a hinged bottom floor a puff here is good too.
2. Standing at the back of the hive, remove the hive roof and gently lay it on the ground.
3. Starting at the back of where the bees are, remove the follower board. You can peer inside the hive and observe the cluster.

Take your time, and make your movements slow and considered. This helps to keep you and your bees calm. PHOTO: *Alphapix.*

This observation window allows the beekeeper to observe what the bees are doing inside the hive without disturbing them.

4. Using your hive tool or knife, insert this between two of the rear top bars and prize the knife/tool to the side, breaking the propolis seal. Remove at least three empty top bars until you reach the start of the bee cluster. Keep these three top bars out of the hive as this space gives you working room to move and manipulate the top bars.
5. Repeating the action of prizing between the two next bars, gently release the propolis on each end of each top bar. Place your fingers under the top bar and gently lift the top bar directly upwards out of the hive.
6. If there is any resistance at all, stop and get your hive tool/knife and from the hive floor, slide the tool up each side of the inside of the hive. This will cut any bridging comb that is anchoring the comb to the inside wall of the hive.
7. When handling the top bar always keep the comb vertical. The minute the comb is moved at any degrees horizontally the comb has the potential to break off the top bar.

8. I balance the underside of the top bar on my index fingers. In this way the comb finds its center of gravity.
9. Once you have inspected this bar, return it to the back of the cluster leaving some space to manipulate the next fame.
10. Remember to always return the comb in the same position and orientation. A good idea is to paint a dot on one end of each bar so you can remember how the comb came out of the hive. If you mix up the order of the frames and put a large frame of brood near the edge of the cluster of bees, the bees may not be able to cover it and keep the brood warm due to their small population and the shape of the brood nest changing. Keep the brood combs together and position the honey and pollen frames on the edges of the hive.
11. You can shake or brush the bees off back into the hive to have a clearer inspection of the cells.

During Sonya's first solo hive inspection she could not locate the queen. After carefully lifting and inspecting each comb and looking at both sides, she was concerned that her new hive had somehow become queenless. When you are new to beekeeping, finding the queen is like looking for a needle in a haystack. I remember in my first year of beekeeping I think I only saw my queen a handful of times. Try not to worry about seeing the queen but rather look for evidence of her.

- Can you see capped worker brood? If capped worker brood is present then the queen would have laid an egg 11 days before.
- Can you see white C-shaped grubs in the base of the cells? If you can see these grubs then a queen was present around six days ago.
- Can you see tiny white eggs in the very bottom of the cells? If they are standing upright on their ends they have very recently been laid by the queen. If they are leaning onto their sides they are nearing three and a half days old and about to turn into young larvae.
- If you look into the bottom of the cells and see only glistening jelly this means that there are very young, four-day-old larvae swimming in a pool of royal jelly.

Queen spotting is an art which will develop over time. Concentrate on the brood to read how your colony is progressing. It is very easy to miss the queen if she is on the side of a comb or busy laying, as her long abdomen will be inside a cell.

Always remove a top bar frame vertically from the hive. If you feel any resistance slide your hive tool up either side of the hive to release any bridging comb. PHOTO: *Alphapix.*

Always hold top bar comb vertically to prevent comb breaking from bar.
PHOTO: *Alphapix.*

If you position yourself so the sun is on your back it makes it easier to view inside the cells. Raise the comb up and down to gain the right angle. PHOTO: *Alphapix.*

When you are inspecting a frame I find it makes it much easier to view the cells if you have the sun at your back and hold the frame upwards so that sunlight comes over your shoulder and shines into the base of the cells. If your eyes are failing have your reading glasses on or use a magnifying glass.

What are you looking at when you inspect a frame?
Whilst you are assessing each frame have a checklist in your head of what you should be seeing as you evaluate the health of your hive. Observe these components:

Comb: Drawn comb is one of the most valuable resources of a hive. Is the comb straight? How old is the comb? The comb darkens after each succession of brood rearing. Darkened comb has a potential to harbor diseases. It is important to cycle out old comb. If you hold empty comb up to the light you should be able to see some light through it, if not mark the bar or frame and aim to recycle it out of the hive. You can do this by moving it to the back of the Top Bar hive or into the honey super and harvest once full of honey. If there is capped brood then the comb should feel heavy with the larvae and capped honey.

Is there space for the colony to expand if your inspection is being conducted in spring and summer?

Healthy Brood: Is the queen laying in a good laying pattern? The brood should fill the frame and there should be only a few empty cells in similar aged brood. The queen lays in a circular pattern beginning in the center of the frame. Are the cappings all uniform in size and color. If there are any cappings with small perforations, are different in color or look greasy, get a matchstick and twist into the cell and remove the larvae and observe.

Worker brood will be found in the center of the comb and drone brood will normally be found along the side and bottom edge of the comb. Expect 15–20 percent of drone brood in natural-drawn comb. Empty drone brood cells are larger in size than worker brood.

If you can see eggs there should only ever be one egg per cell unless the queen is very young and has just started laying.

Nectar and Honey: Honey is stored on its own entire comb face or as an arc at the very top of a brood frame. Honey is often stored in drone brood

This photo shows capped worker brood on the far right and uncapped brood. These look like large white c-shaped grubs. On the left are a few cells with freshly laid eggs. You can tell they are freshly laid as they are standing upright. By day three they will start to lay on their side and change into tiny grubs. Observing this frame you know that the queen has very recently been here laying and the brood looks very healthy.

Eric conducting a hive inspection during a warm, still day in his urban garden.

This naturally drawn comb has capped honey and nectar which is still being filled into the cells. When the nectar has been ripened, the individual cell will be capped with a thin layer of wax.

This is a good example of a healthy frame. There is a thick arc of capped honey at the very top of the frame, followed by a thin arc of pollen directly under the honey, followed by a good pattern of worker brood.

It is normal to observe some queen cups along the sides of naturally built comb. These do not mean your colony is about to swarm. The hive will naturally construct these in case of emergencies.
PHOTO: *Alphapix.*

Healthy-looking larvae attended by healthy-looking worker bees.
PHOTO: *Alphapix.*

This natural comb shows a section of capped worker brood and stored bee bread in nearby cells.
PHOTO: *Alphapix.*

comb. Nectar will be uncapped and can be quite watery. If you see nectar in the combs then you know that a nectar flow is happening in your area. It is estimated that it takes one cell of honey and one cell of pollen to raise one bee.

Pollen: Pollen is often stored just below the arc of honey towards the top of the comb. Pollen cells are always only three-quarters filled. Pollen can be many different colors depending on its source. Pollen is very important as a protein source for brood rearing.

Queen cups/cells: Queen cups are common and often built on the side margins of the comb. They are constructed in case of emergencies. They generally do not mean that the colony is about to swarm. If you tip the comb up and look into the cup it should be empty. It is best to leave them alone. If you see queen cells along the bottom of the comb then this does mean that the colony is preparing to swarm.

Health of bees: Are the emerging bees healthy? Do any have any deformed crinkly wings? This could be because there is a high level of varroa mites in the hive. Open an older-looking drone brood cap with the corner of your knife or hive tool. Drag the larvae out and check for varroa on it or in the cells. Varroa prefer drone brood due to the longer time they spend in the cell so varroa mites are most evident in them.

Are there food stores in the hive for the bees? Is the colony stronger or weaker than the last inspection? There should be plentiful bees covering the face of each comb of brood. Can you see any signs of disease within the brood?

Sound: Are the bees making a gentle constant buzz as you work inside the hive? This is normal. If you open the hive and are met with a "roaring" sound, which is deeper and more constant than the usual sounding buzz, your hive may be queenless. This roaring sound is created by bees fanning whilst exposing their Nasonov gland. Bees will do this when a queen is absent to try to guide her back.

Smell: A healthy hive smells of warmed wax and the sweet aroma of honey. If you smell a fishy smell or a fermenting aroma investigate further. A fishy smell is often associated with American foulbrood. If your bees have been poisoned, a large number of bees could be on the hive floor and can emit a fermenting odor.

The Queen: Is she moving normally around the face of the comb? Is her abdomen enlarged and long (which means she is fertile and full of eggs)? Are there hive bees attending to her? Often you will see hive bees create a round flower petal pattern enclosing the queen as they feed her and groom her and spread the queen pheromones from her to the hive. Large solid areas of sealed worker brood and concentric rings of eggs and larvae at similar stages of development are a sign that you have a good queen.

This worker bee is taking a break a cleaning up some spilt nectar on the top of a frame. Notice her long tongue.

Important nectar and pollen sources for spring bees
- Alder
- Apple species
- Black locust
- Cherry and peach species
- Chickweed
- Clover
- Dandelion
- Elderberry
- Eucalyptus
- Honey locust
- Linden tree
- Quince
- Robinia

- Shadbush (Amelanchier)
- Tulip popular
- Willow, especially pussy willow

Late spring–early summer: building up
During this time a colony will build up rapidly. The colony requires space for brood rearing and food storage and also space for the new adult bees.

It is important to recognize when a honey flow is in progress in your area as this food will give your bees the resources they need to construct wax quickly. If a hive becomes cramped and runs out of room for the queen to lay in or for the colony to store honey they will start the process of swarming. It is important for the beekeeper to stay one step ahead and provide extra space regularly.

Colonies with small populations emphasize brood production over honey storage. These colonies will use most of their incoming food resources to feed and nurture brood and build new comb to provide a larger nursery area. They will have a high brood to adult bee ratio. These adult bees are required to care for the brood and also to keep the capped brood warm until it emerges.

What is a nectar flow and how to tell if one is on in your area?
A nectar flow, sometimes called a honey flow, is when one or more major nectar sources are blooming in your area and the weather is warm and settled, permitting the bees to forage and collect this resource. A nectar flow is the main opportunity for the bees to collect honey to sustain them through winter. A nectar flow lasts for around two to three weeks in most areas and is seen in the beehive by surplus honey.

If you observe that your combs have brood and nectar filled cells side by side together in the center of the comb, your bees may be running out of space and are not able to provide separate space for the queen to lay eggs and the worker bees to store honey in. Give your hive extra empty frames either side of the brood area (or above or below the brood area depending on your hive design) so they can build this out and use it for food storage. Having honey and brood stores scattered through a face of the comb can lead to the colony becoming honey bound and the hive swarming.

When a particular area has its main nectar will depend on the natural flora growing in that area and the recent weather patterns. Asking experienced local beekeepers is a good way to learn when to expect this in your area.

Weather is pivotal to a good nectar flow. Plants require regular water and sunlight to produce sugar through photosynthesis. Healthy vigorous plants will produce more nectar than plants stressed by too little water or an unseasonably cold temperature. Warm days and cooler nights lead to a higher secretion of nectar. Of course to be able to collect this nectar the bees require settled weather so they can leave the hive.

Signs that a nectar flow is happening in your area

- The nectar flow typically occurs during the summer months and lasts for one to three weeks.
- There are many flowering trees and plants in your area. You may walk under a tree and hear bees working en masse and the sweet smell of nectar in the air.
- Your bees will appear content and very busy when you open the hive. They will often take little heed of you.
- You will see capped honey and open cells containing nectar in many of the frames.
- Your bees may be busy building new comb as they have the food resources to undertake this taxing role.
- When you stand near your hive you will be able to smell the sweet scent of honey.
- At night you may see your bees "bearding". This is when a large proportion of adult bees hang in a curtain from the hive entrance. They do this to lower the level of humidity within the hive as the action of ripening nectar to turn it into honey produces a lot of moisture. These bearding bees will also fan the entrance with their wings helping to direct humid air out of the hive.

If you look down between frames of your Langstroth, Warré or Flow hive and observe lots of new white comb you can be assured that a nectar flow is present in your area. Bees require a high level of nutrition from nectar or honey to build new comb.

The bees are building new comb to either side of the original combs. New comb is white and very fragile and needs to be handled with great respect. Propolis can be seen on the side of the top bar.

- If you remove a frame and shake it and nectar is scattered out then this is nectar that has been collected today.
- When you open up your Langstroth/Flow hive and find fresh white bur comb between the top bars and the crown board, the bees are said to be white waxing, which is a clear sign that a nectar flow is on.

Sonya and her Top Bar hive

Through spring and into early summer Sonya chose to continue to feed her young nucleus colony a sugar syrup solution to supplement the available forage. By early summer her bees had stopped feeding off the syrup and were bringing in nectar. It is probably of no surprise that the bees prefer their natural food to white sugar, if they can find it. In fact, feeding of sugar syrup decreases an adult bee's life span as the bee uses body fats to convert the sugar to honey.

Sonya's Top Bar hive has the entrance holes in the middle so her bees commenced building new comb to either side of the initial four nucleus frames. During this time of comb building it is very important that the new comb is assessed regularly to ensure that the bees are building straight comb on individual top bars. This is because this comb will form the building blocks of the brood area and it is imperative that these combs can be removed bar by bar to be inspected for exotic diseases. Not only is this important but it is a legal requirement in many countries, including New Zealand.

A good management technique to get in the habit of doing is to place an empty top bar between the last and the second to last bar for partly built top bar combs. The newest comb, being built on the empty bar is between two already constructed or partly constructed combs so these provide a template and encourage straight comb building.

Sonya checked her hive weekly during this time. This check entailed opening the hive roof and moving the follower board away and observing the end combs and perhaps introducing an empty top bar between the two last top bars being built. Sonya could do this very quickly without needing to wear protective gear or use smoke, as the bees were often not even aware of her presence.

During the heavy early summer nectar flow Sonya's bees were building a full new comb every week. Regular rainfall between warm settled sunny days ensured that plants were producing ample flowers full of nectar.

Eric and his Warré hive

Eric's bees are building up very quickly. Within four weeks his swarm of bees have built new comb and were raising brood in the Warré hive, which had grown to three boxes high. Swarms are known to grow very fast and this is because these small units are made up of predominately worker bees aged between 12 and 18 days. These aged bees can produce wax from their specialized gland. Swarms have access to higher levels of queen substance and other pheromones, which drive them to work very hard at comb building, foraging and food storage.

Eric's hive is also positioned in the middle of a suburban area where there is ample forage from neighboring gardens and parks. It is a well-known fact that urban bees are much more productive than rural bees as they have such a varied and plentiful pollen and nectar sources. The warmer urban temperature caused by the "heat sink" of buildings also allows plants a longer growing season.

> Eric wants to increase his hives to three for his large suburban garden. It is always a good idea to have at least two hives for the simple reason if one hive fails, becomes queenless or requires more food stores you can easily transfer frames from the other hive to it. Having two hives also does help with your learning as they will never be the same and you can easily learn to assess differences in strong and weaker hives.
>
> If Eric did not want to create a new hive he would need to continue to add boxes to the bottom of the Warré hive, in the process of nadiring. This would provide more room for the bees to move down into, thus preventing the colony from becoming cramped and attempting to swarm. By splitting his hive in early summer Eric is happy to forgo ample honey harvesting, as these divided hives will require all their resources going into winter.

Dividing hives is the best way to propagate your hives. It is normally done in spring to help prevent swarming. Studies have shown that it is possible to make splits when the swarming tendency is not present but the hive will raise a stronger queen when the swarming impulse is high. (Cowder and Harrell, 2012). The reason for this is that when pollen and nectar sources are high the colony can raise a strong, healthy and well-fed queen.

The recent Bee Informed Survey in the United States discovered that when beekeepers permitted their colonies to split using a queen cell or to rear a new replacement queen, they lost fewer overwintering colonies than beekeepers who introduced a mated or virgin caged queen (beeinformed.org/results/the-bee-informed-partnership-national-management-survey-2014-2015).

To split a hive you do require the hive to be strong and be headed by a healthy young queen. There needs to be at least eight combs of brood before a hive can be divided. Ideally you would split a hive when you see uncapped swarm cells. Using a swarm cell or a bought queen cell gives you a higher possibility of ending with a quality queen.

Using the "emergency" method below, the bees are forced to raise a queen using an egg in a worker cell and there is a chance that the queen will be somewhat compromised because it hatched from a smaller-sized cell. At the worst you can end up with an intercaste queen. An intercaste queen is born from worker larvae which was really too old to be changed into a queen generally because they did not receive royal jelly from a very young age. Intercastes have both queen and worker characteristics.

Within two hours the queenless split colony realizes that they are queenless due to the drop in queen pheromones. The worker bees immediately begin to enlarge some of the cells that contain eggs and very young larvae. The bees may start raising a single queen but more often will raise several cells. These cells are known as emergency queen cells.

Emergency cells are different from swarm cells, which are built at the bottom edge of the comb. Emergency cells derive from worker cells so appear on the face of the comb. As the worker cell faces outwards the orientation of the cells has to be drawn out and around to face downwards. This usually means that some of the worker cells below this cell are destroyed to make way for the elongated shape of the queen cell.

Eric systematically conducting an inspection through this Warré hive.

The developing larva of this emergency cell is fed a diet that consists of only royal jelly. After eight days the cells are capped and the larva pupates into a new queen and hatches. The queens fight it out until the victor is found. The new queen conducts her mating flights and then commences heading up this new colony of bees. This whole process takes around 14 days, a few days shorter than the normal 16 days, as the bees have started with a young larva.

This is the process Eric followed to make two Warré hives from his one using the emergency queen procedure:
1. Check hive to split and ensure that there are frames of eggs and very young larvae present, as these will be used to make a new emergency queen.
2. Assemble all the equipment and hive equipment that will be needed to set up a complete new hive. Set up this hive in another area of the garden. It is best if this hive is located as far away from the parent hive as possible.

3. Go through the parent hive and locate the queen on a frame. Check this frame and ensure that there are no swarm cells with larvae or eggs in them. If these swarm cells are moved over with the queen, this new hive may swarm leaving the hive even more depleted of an adult bee population.

4. Transfer the queen on this frame into a new hive box. Have this hive box right next to where you are working so that there is no risk of the queen falling off the frame while being transferred.

5. Once you have transferred the frame with the queen, go through the remaining frames and choose frames that mostly have capped worker brood. This will provide this young hive with a population of hive bees to look after the queen and will soon mature into foraging bees.

6. Choose some frames of pollen and capped honey. Always check each frame thoroughly for a queen as some hives will have two queens. This new hive will lose all its foraging force of bees so will need resources to get them through the initial period of no foraging bees. The nurse bees will require honey to feed themselves while they look after the queen and larvae until this larvae has hatched as new hive bees. Position the frames of pollen and honey on the sides of the brood frames. Place the pollen frame near the entrance if your hive design allows. Do not divide the brood nest.

7. Choose a further three frames which are covered with bees. Hold these over the new hive and shake the bees into the new hive one by one. This introduces hive bees into the new hive to look after the queen and keep the brood warm.

8. Move the new hive body with the queen, brood combs and hive bees to its new position in the garden and seal with hive roof. You will see some of the flying bees from this new hive return to the parent hive, as this is where they have orientated.

9. In the hive that is now queenless, you need well-fed nurse bees which can produce and feed copious amounts of royal jelly to the newly hatched larvae to turn them into queen larvae. A way to encourage royal jelly production is to mash up pollen and honey cells on a frame. Use your hive tool to break the cells and mix the pollen and honey into a sticky mess on the face of the comb. This mix of food will get the nurse bees well nourished and able to produce royal jelly to feed the chosen larvae which are being raised as new queens.

10. In exactly five days check the queenless hive. If there are a variety of queen cells in different stages of building, remove any of the larger more

developed queen cells as these have been developed from larvae which were on the cusp of being too old and will not make a good queen. As these cells are older than the desired very young larvae, they will hatch first, emerge as new virgin queens and destroy all the other developing queen cells.
11. Do not disturb this hive for at least three weeks. This gives the hive an undisturbed environment to raise a new queen and for her to take her virgin mating flights.
12. Check both hives in three to four weeks' time to see if they both have laying queens present.

Can good-quality queens be raised by this emergency method?

The bees are in crisis mode when they are raising emergency queen cells. If they fail to raise a queen the colony will perish. The bees need to create almost a right angle to the shape of the cell to make the outward facing cell turn into a long hanging vertical cell. To get the chosen larvae into the base of the vertical queen cell they need to float it out on a waterbed of royal jelly. To do this they need to thin the royal jelly (it is thought that this can cause it to lack its correct nutritional value). The bees also often make multiple emergency cells so these vital resources are overextended. To produce abundant royal jelly the bees need access to copious food resources. If this is happening in early spring or in autumn, nectar and pollen may be hard to find. If the frame is older comb it makes it difficult for the bees to build the cells downwards and destroy the cells under the queen cell, as it has been toughened with layers of old cocoons. To ensure success using this method of raising a new queen:

- Carry out this process when a strong nectar flow is happening
- You may choose to feed the bees additional sugar syrup
- Choose open, very young brood on clean, drawn fresh comb, not dark, hard old comb
- Do not disturb the colony for at least three weeks
- If after three weeks the colony still has no new queen, add another frame of eggs and very young larvae and repeat the process
- A healthy, well-fed and raised queen larva should be enclosed in a queen cell which weighs at least 1 oz (25 g).

Sarah taking possession of her new nucleus hive in her garden.

Sarah and her Flow hive

It was late spring before bees were introduced into Sarah's new Flow hive as she had to wait to receive her new hive as a kitset from overseas. Her Flow hive is an eight-frame, full-depth super with a honey super, which contains six of the thicker flow frames. Eight-frame Langstroth hives are rarely used in New Zealand, the ten-frame Langstroth being much more common.

Her hive came flat-packed all the way from Oregon, USA where it was made from cedar. With the post, customs and tax requirements, the package was a very expensive parcel. The clear instructions made assembling the hive a breeze. A cordless drill, a set square, some wood glue, a screwdriver and a hammer were required. It took around two hours to get all the pieces together but very soon Sarah was the proud owner of a new hive sitting proudly on her kitchen floor.

To give the wood some natural, low-toxic protection Sarah painted the outside of the hive boxes and roof with tung oil. This is a natural oil which is very good at wood preservation. If you can't find this type of oil, use raw linseed. Oiling the wood is important for wet climates.

Sarah has decided to provide empty Langstroth frames without wax or plastic foundation

With a few common tools it does not take long for Sarah to assemble her new hive.

Sarah used tung oil, a natural oil, to weatherproof the cedar wood of her Flow hive. Raw linseed oil is another option.

Using a hive tool Sarah carefully prized each frame out of the nucleus colony and moved them into the brood box of the Flow hive. Sarah ensured that each frame was placed in the same orientation and position as it came out of the nucleus.

Sarah and her newly installed Flow hive. Sarah has elevated her hive on a wooden pallet and some concrete blocks towards the back of her garden. She has left some space at the back and sides of the hive to give her working room when she is in the hive.

to allow the bees to build their own unique wax to their own dimensions. To ensure that the bees build straight comb from the top to the bottom of each frame Sarah has inserted long pieces of guide wood into the groove on the underside of the top bar of the wooden frame. These guides come in the hive package.

Sarah has not bothered wiring the frames. This is for a couple of important reasons. Firstly as these frames are being placed into the brood area they will not be used in a centrifuge to extract honey. Wire is often inserted to give the comb strength when it is being spun at high revolutions in electrical-powered honey spinners and also to embed the wax foundation so the wax does not slump in the heat. Wiring frames does require some extra pieces of equipment and of course extra time and expertise. Lastly, and perhaps most importantly, recent research conducted by a student from the University of North Carolina discovered that all the empty brood cells were located along the metal wires in hives she was studying. The researcher began testing younger brood positioned next to the wires and found that they had six times more iron in them than those that weren't located along the wire (newsandfeatures.uncg.edu/kaira-wagoner-honey-bee-research). As a result of this research I personally think it is prudent to not use steel wire in any part of the beehive.

When positioning a flow hive in its permanent position it is recommended to have the hive leaning backwards around 3–4 degrees. This will make honey harvesting easier as it allows the honey to drain out the back through the flow frames. Once positioned in her garden, Sarah carefully transferred the nucleus hive into the lower brood box of her new Flow hive. Sarah did this by removing all the empty frames for her new hive box and then carefully transferring each frame with clinging bees into the box. She made sure she kept the same orientation and position as how they came out of the nucleus box. Placing these four frames in the center of the brood box, she enclosed them with empty frames.

The bees slowly but steadily began expanding from the full-depth, four-frame nucleus colony. The bees slowly moved out from the center cluster to the new empty frames which enclosed either side of the cluster. Sarah could witness lots of bees bringing pollen back into the hive. This is a clear indication that the queen is busy laying, as pollen is used to feed the developing brood.

The bees quickly built out all but the two outer frames. To encourage the bees to build out the lower box completely, Sarah swapped frame two with frame one and frame seven with frame eight so that the bees could develop the maximum space to raise brood and store food.

The bees will build out the foundation comb commencing at the top. They build out the hexagon sides so that the cell is formed and the queen can lay an egg at the base.

Late summer–early autumn

Some summer nectar sources for bees
- Basil
- Beans
- Buckwheat
- Chestnut
- Clover
- Cucumber
- Globe thistle
- Mint
- Persimmon
- Privet
- Pumpkin
- Queen Anne's lace
- Sunflowers
- Thyme

By late summer the main nectar flows are over in most areas. A hive in its first season has spent all its resources in building comb, raising brood and storing honey to last it through the food-scarce winter. Do not expect a sizeable honey harvest, if any at all. I always believe it pays to err on the side of caution and not harvest any honey at the end of your new hive's first season. If you leave all the honey for the bees for winter and they still have available stores come spring, then you can harvest a comb after seeing your bees survive the winter on the best food available to them, rather than unnatural white sugar.

Robbing

Robbing becomes an issue for all hives during this time of the season. As nectar sources start to become sporadic or scarce, bees seek out other sources of food to gather to sustain their colony through winter. Smaller or weaker hives are under constant pressure from attack from bees from other colonies and also from wasps. Robbing is a form of foraging when the foraging bees will seek out honey from a weaker hive and steal it. This only occurs when natural nectar sources are scarce, as bees prefer to visit flowers.

Robbing behavior has a distinctive pattern of flight. The robber bees fly in a zigzag pattern and hover around the entry. These robber bees will also look for alternative entrances such as cracks in the corner of boxes or under the hive lid. Robbing shouldn't be confused with drone flights or orientation flights of newly matured foraging bees. These bees will fly in a slower figure of eight pattern facing the hive. Robber bees' flight is similar but in more of a zipping or scurrying pattern. During a robbing session you may see bees fighting on the ground in front of the hive or at the entrance. Robbing bees can remove all the honey stores and the colony under attack can die from the heat stress due to elevated temperatures created within the hive.

If you have a screened bottom board and observe pieces of wax yellow in color covering the sticky board during a time when there is a lack of a nectar flow, this can indicate a massive robbing incident. The frantic robber bees chew the wax cappings as they remove the honey. If you open a hive and find fewer bees and all the capped honey empty, with the cappings looking like they have been chewed, this is often an indication that your colony has been robbed out. Without artificial feeding this hive will not survive the winter months without any food stores.

You can help to prevent robbing by only opening hives when absolutely necessary and keeping them open for the shortest time possible. Keep all your hives in good condition with no cracks or holes. With a small colony, reduce the hive entrance using a strip of wood with a Langstroth/Flow hive or Warré hive, or some corks with your Top Bar hive. Leave a $1\frac{1}{2}$ in (4 cm) gap and observe if the small colony can defend this.

A robbing screen is another option and can be easily made at home. I use some fine aluminum mesh that I buy from a hardware store. Cut it into a strip large enough to cover the hive entrance. Using a stapler, I fix the top and bottom of the mesh to the hive with staples but leave both sides open. Resident bees will quickly learn to crawl out and along to get out from behind the screen. Robber bees and wasps tend to hover in front of the entrance when attacking so can't penetrate through the screen. All small colonies will benefit from these screens but I place them on all my hives as wasps are a common pest in my area.

Wasps do not travel far so if you observe lots of wasps around your hives try and follow their flight path back to their colony. You can kill the wasp colony by pouring a cup of petrol down their entrance hole and covering with something solid. Do this at night and wear your bee suit as a precaution.

This top bar of sealed honey is appropriately prepared by the bees for you to harvest.

After choosing a frame of ripe capped honey to harvest, gently brush the bees off the frame over the hive so they fall back into the hive. Quickly cut and place the honeycomb into a covered container.

Sonya's harvested honey straight from the hive and still warm!

Natural comb honey placed in glass jars and ready for some runny honey to surround it.

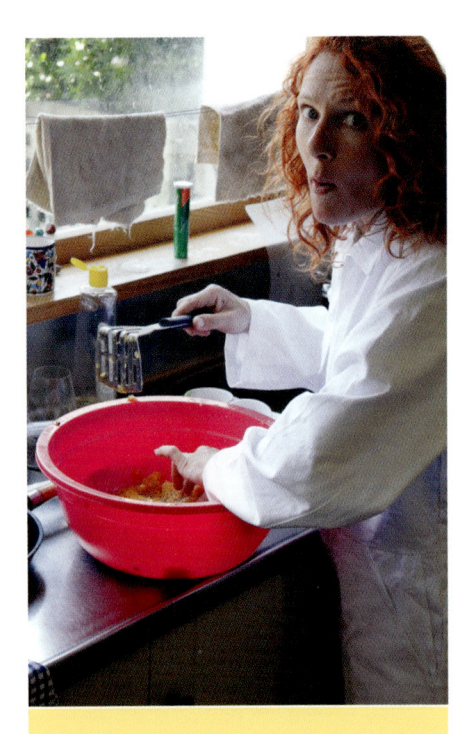

After cutting up some comb honey and placing it in glass jars, Sonya mashed the remaining honey using a potato masher. After straining it through a kitchen sieve, Sonya is left with runny honey to pour over the comb honey.

Sonya and her first jar of organically managed, bee focused, backyard honey!

If you witness a robbing session, what can you do?
- If you know that one of your other hives is doing the robbing you can remove their hive lid. This will cause the robbing bees to return to their hive to defend their own stores.
- Quickly reduce the entrances of the hive being robbed, or any gaps where the robbing bees may be gaining access.
- Cover the hive being robbed with a cotton sheet and turn on a water sprinkler to mimic rainfall. This will encourage the robbing bees to return to their own hives.
- Close all entrances to the hive being robbed.

Extracting honey

Sonya and her Top Bar hive

Sonya's Top Bar hive has gone from strength to strength over the season. The four-frame nuc she started with in early spring, headed up by a young queen, has now built up to twenty-six frames of naturally drawn wax filled with pollen, brood or honey. As her colony expands the bees start storing honey in the bars furthest from the entrance and the brood cluster. These combs are filled with nectar and then this nectar is turned into honey via a process of evaporation. When the water content is less than 19 percent the bees will cap the honey with a thin layer of new wax.

As Sonya's Top Bar bees move into winter, as a clustered unit they will move along the bars to reach this food store to nourish them over the cold winter months (when not many plants are flowering and it is often too cold to leave the hive).

Sonya wishes to harvest one comb just to taste her first honey from her bees. To harvest honey, Sonya assembles the following equipment:
- Bee brush or large feather to brush bees off comb
- Bee suit and gloves
- Smoker
- Knife
- Large lidded container.

This is the process Sonya followed to harvest her first comb of honey:

Sonya gently opened up one end of the hive. As Sonya's hive design had the entrance holes in the middle of the long side, the bees will store excess honey at either end of the hive. The brood will be around the entry holes. Carefully sliding a knife upwards along the inside of the hive, she cut any comb attached to the side of the hive. Levering the knife under the top bar to break away any propolis seal and then gently lifting the top bar straight up and out, Sonya could assess each bar. It is very important to always hold the bar so the comb is vertical to the ground. If you swing it horizontally it can easily break away from the bar. It is surprising how heavy a comb of fully capped honey is! What Sonya is looking for is a bar full of capped honey. If you harvest uncapped honey it can quickly ferment, as its water content is too high. It should look white. The first two bars she assessed had new comb or comb filled with nectar but not yet capped. She returned these to the hive in their original position. The combs nearer the brood area were fully capped and full of honey as these would have been older and had already been completed as the brood expanded.

Gently brushing the bees off the comb, Sonya then quickly cut the honeycomb off the top bar, leaving about 1¼ in (3 cm) of comb along the top bar. Replace the bar outside of the other bars that have comb. The bees will eat the remaining honey and start rebuilding the comb. Sonya used a large plastic container with a tea towel to form a lid to place the honey in.

It is a good idea to do this in the morning around 10am. At this time most of the forager bees are out and about but it is cool enough for the combs to be firm and the honey is not as runny as in the middle of a hot day. It can be a sticky process nevertheless. Disaster can and does happen. It is very easy for the fragile comb to break off the top bar and it is only with practice that you will get better at removing and checking the combs. If a comb does break and has developing brood, gently prop it up between two hanging combs. The brood will still be fed and nurtured and when all the brood has hatched you can remove the empty comb from the hive. The new white comb is especially fragile but becomes tougher as it ages and turns a rusty color.

After closing the hive, Sonya took her bar of harvested capped honey into her kitchen. After closing all the windows to stop any flying bees being attracted to the smell of honey, she used a kitchen knife to cut up the large honeycombs and placed them in glass jars. With the remaining combs she used a potato masher to smash the wax cappings and cells to release the honey. It is then a simple process of filtering this mushed honey through a sieve into another lidded container. A fully capped top bar comb will normally yield around four medium jars of honey.

The sticky wax left in the sieve can be gently washed in warm running water and allowed to dry on some paper towels.

Eric and his Warré hive

Eric had made the decision to split his hives to increase his number from one to two. This is a good practice. If you have two hives you can use the hives to evaluate and compare with each other. If one hive is not doing well, brood can be moved from the stronger hive to the weaker one. If a hive was to become queenless for any reason it is much easier to transfer a frame of appropriate-aged larvae from your other hive rather than trying to source a queen from a breeder in short time.

As Eric's bees have been extended, building natural comb from scratch and developing colonies that have a healthy population and adequate stores, he accepts that he may not

Eric uses a hot knife to carefully slice off the wax capping of each frame of comb honey before placing in honey extractor.

This small honey extractor is operated by hand so it is very easy to adjust the speed at which the honey is spun out of the cells by centrifugal force.

have any excess honey to harvest this year. He wants to keep his bees as naturally as possible and does not want to feed them white sugar syrup if he can avoid this. In saying this, he is finding that the bees are bringing in copious amounts of nectar, which is often the case when you are an urban beekeeper.

Eric is finding that cockroaches are living in the coarse sawdust in the quilt box. Cockroaches are common in dry and hot weather. Of course you cannot use any type of insecticide so close to the bees so Eric has come up with a simple solution. He places a natural cockroach trap on top of the sawdust. This trap contains a surface that is ultra-sticky. When a cockroach walks across the surface they are stuck fast.

Avoid wearing aftershave or perfume around your bees. Eric discovered that his bees do not appreciate the smell of sandalwood oil. After using this as a moisturizer he was quickly stung by two of his normally docile bees when he ventured out near his hives.

Eric prefers to consume his honey as natural comb honey. Comb honey is most attractive when it has been built on new, clean and white wax. The Warré hives design follows the theory that the bees naturally build down from the top of the hive cavity. Because of this, comb used to store honey has undergone at least one cycle of brood raising. This means the comb will have some discoloration and be a darker brown color rather than the clear white color. With this in mind, Eric has decided to manage his hive in a Langstroth/Warré hybrid system. He has contained the brood area in the lower two boxes and restricted the queen to this area by using a queen excluder screen. A queen excluder is a metal or plastic grille which has spacings that allow worker bees to squeeze through but does not allow the queen, with her wider abdomen, through.

Above the queen excluder Eric has placed two further boxes which the bees will use to store honey. As the queen is unable to reach this area the comb remains pale as no brood rearing is done within the cells. Comb darkens with each cycle of brood rearing because when the pupae hatches they leave behind a very thin film of cocoon coating.

By late summer Eric's bees have built out the upper two boxes with comb and are in the process of filling each and every cell with sweet nectar, ripening this by evaporating the excess moisture and capping the honey with a very thin layer of wax. This preserves the honey so it can be used during winter when sources of food are hard to come by and the low temperatures outside the hive are not conducive to flying.

Realizing that he could harvest some honey in this first year Eric bought a small honey extractor and with careful use he extracted 30 lb (14 kg) of honey, along with many different chunks of delicious honeycomb. Because there is no foundation (including supporting wires) in his Warré frames, he had to extract very carefully and not spin the frames too fast. Two frames were obliterated at first until he understood just how fast to operate the spinner. For uncapping he used two large kitchen knives, one in a jug of hot water while he was using the other to take the wax capping off the frames. Next season he intends to invest in an electric uncapping knife. This should be more efficient and cause considerably less damage to the comb in the frames as well. He also discovered that it is very important to harvest your honey in a sealed room. He started the process outside but had to retreat quickly inside after the bees started to reclaim their honey!

Sarah and her Flow hive

As Sarah's hive was started so late in the season (early summer) she has no expectations of a honey harvest. Furthermore she wants to ensure that her bees are set up with ample food stores to see them go into winter. Sarah has continued to feed her bees a sugar syrup solution and the bees have built out all the foundationless frames in the brood box. The natural comb is straight and firmly attached to the top and side bars of each frame.

Sarah did discover that the bees were not going up through the queen excluder onto the flow frames. She placed the in-hive feeder up in the Flow frame box and took away the queen excluder for a short time to encourage the bees up onto these new frames to start

A non-toxic cockroach trap on top of the quilt layer on a Warré hive. Any cockroaches that walk across the sticky surface will get trapped.

This queen excluder is designed to allow the smaller worker bees access into the upper honey box but not the queen. Many beekeepers call it a "bee excluder" as most bees are hesitant to pass through it.

building them out. Perhaps as the frames are brand-new they did not smell attractive to the bees. This is a common problem with all plastic frames. It is recommended to paint all plastic frames including the Flow frames with some melted beeswax to encourage the bees onto them. By removing the queen excluder there is always a risk the queen will go up onto the flow frames and lay in the cells. As the cells are so large she would only lay drone brood. This is the risk when the queen excluder is removed.

During her last inspection she was unable to insert the final frame back in the brood box. She had received a number of stings and was concerned about squashing more bees and making them even madder. She had left her eight-frame brood box with only seven frames. The bees, having recognized this extra space, had quickly filled it with natural "crazy comb" (brace comb) which spanned between the end frame and the inside edge of the box. It is important to remove this as this comb cannot be removed to be inspected, a legal requirement in many countries. It can be difficult to insert that last frame but by lining the frame up properly and very slowly pushing it down to allow the bees to move out of the way you can avoid bee losses but maintain the all-important "bee space".

The top Flow frame's box also has the in-hive feeder. Sarah found that by taking the queen excluder off for a short time, the bees were encouraged to come up onto the frames and start building them out with new wax.

This is a good frame of food stores to feed the bees over winter. This frame will be placed to one side of the brood cluster.

This is brace comb or crazy comb which has been built by the bees when a too large space was left in the hive. Sarah gently smoked the bees off this comb and cut it away and removed it from the hive. It is important to prevent this comb being built as it will have to be removed for a full hive inspection.

During an autumn check it is important to check the health of the brood, look for any signs of varroa infestation or exotic diseases and check that the hive is still "queen right" and has a laying queen.

Bee space is the measurement required for a bee to pass between two structures. It is generally ⅜ in (7.5 mm). If the space is too small the bees will fill this space with propolis, if the space is too large they will build comb.

The bees use the frames next to each end of the hive for storage of pollen and honey. The bees choose to use these frames for pollen and honey storage as these end frames are more prone to temperature fluctuations. The central frames are used by the queen as a nursery, where the brood is cared for. Being in the center of the hive and covered with the majority of the bee cluster, these brood frames are kept at an even hive temperature of around 95°F (35°C).

As a pre-winter check, Sarah assesses all the frames to ensure that the queen is laying and there is enough food to get through the winter. Not finding the queen on the first two frames, these are carefully propped up against the hive as she works her way through the frames. There is plenty of stored honey and pollen and Sarah knows that the queen must be present as she can see eggs and young larvae. Not finding the queen on the frames in the hive, she commences to carefully replace the frames which have been out of the hive only to discover the queen on one of these frames. She was lucky that the queen had not fallen off the frame or been squashed. Very carefully this frame was returned to the hive along with the queen.

She arranged the frames so that the brood frames were in the center of the hive, enclosed by the frames of capped and uncapped honey and pollen. She has chosen to leave the flow frame box on top of the brood box for winter, without the queen excluder.

To help prevent any robbing from wasps or other bees, Sarah reduced the entrance space to her hive by placing a piece of wood in front of the entrance thus reducing the entrance to around 1¼ in (3 cm).

How to harvest honey from a Flow hive

It is important to open your Flow frame super and actually pull out some of the frames to check that the majority of the cells are full and have been capped. This is to avoid harvesting unripe honey, which will quickly ferment. As you become more experienced with your bees and the nectar flows of your area, you may be able to adequately assess the amount of honey by viewing through the window of the Flow frame box.

The queen was located on one of the frames propped up against the hive during the inspection. Luckily no harm came to her.

After any hive manipulation it is common to see the bees hanging off the front of the hive. After a time they will settle and re-enter the hive and get on with their chores.

1. Gather all the required equipment, which includes:
   - The Flow frame key
   - Some long nose pliers
   - A clean, dry container with a lid to hold the honey (each full Flow frame will contain around 70 fl oz (2 liters) of honey)
   - Collection tube
   - Food-grade tube to act as an extension tube from the collection tube to the container (if you wish)
   - Clean tea towel
   - Netting to cover the container to stop debris or insects getting into the honey.
2. Standing at the back of the hive, remove the wooden covers.
3. Using the pliers, remove the top and bottom plastic caps from the frame you wish to harvest.
4. Insert the collection tube with the extension tubing into the bottom slot and arrange the honey container and mesh cover.

Using some long-nosed pliers, remove the caps from the top and bottom of the flow frame. PHOTO: *Alphapix.*

The honey is draining into the container. The hive bees are blissfully unaware of what is going on. PHOTO: *Alphapix.*

5. Insert Flow key into the lower slot at the top of the Flow frame, right through the frame all the way towards the front of the hive. If it is difficult to turn the key, and often it is, only insert the key ⅓ of the way in, turn and then insert it further and turn until you have cracked all the cells open.
6. Turn the key a half turn. It can take a lot of pressure to break the wax seals. Keep a strong pressure on the key rather than a rough jerking action, which could rock the hive and disturb the bees.
7. The honey should begin to flow out the bottom tube in a consistent and strong flow. If it doesn't, repeat the above stage. Cover the top of the jar with the tea towel to prevent robbing by other bees or insects.
8. Depending on the thickness of the honey and the daytime temperature, the honey will take anywhere from 20 minutes to overnight to drain completely.
9. You don't have to fully drain the frame. Any remaining honey will be taken by the bees to refill the cells.

As the honey starts to flow, cover the mesh top with a tea towel to help prevent the scent of honey escaping and alerting robbing bees. PHOTO: *Alphapix*.

It can take a good amount of pressure to crack the wax cells so that the honey will flow down to the collection tube. Avoid any jerky movements. PHOTO: *Alphapix*.

This liter of wonderful, local and organic honey was collected from one Flow frame and took 20 minutes to drain. It was harvested during the prime robbing season of early autumn with no incident of robbing or even bees flying around the back of the hive. I did not feel I needed to wear any protective gear. A tight lid was placed on the jar to prevent excess moisture entering the honey. No further filtering or processing was required. This raw honey is ready to enjoy. PHOTO: *Alphapix*.

This frame of top bar honey is ready to be harvested. All the nectar has been ripened into honey and capped with a thin layer of wax.

Closing the Flow frame:
1. Insert key into the upper slot and turn the key a half turn downwards to close the cells. Remove the key by twisting it back to horizontal.
2. Remove the honey container and replace all the plastic caps on the Flow frame.
3. Wipe away any honey spills on the outside with a wet cloth.
4. Replace the wooden covers.

Why is it important to only harvest honey that has been capped?
Properly harvested and stored honey is the one food that will not spoil and can be safely eaten decades after being harvested. However, there are several important conditions which need to be followed to harvest honey that will have a long shelf life.

Honey will absorb water from its surroundings so it is important to store honey in an airtight container. The hive bees will cure honey with a water content of 17–19 percent. This honey is protected from decay by its high sugar content, thick viscosity, acidity and enzymes.

If a frame of honey is harvested but there is still a large amount of uncapped nectar, the water content of the harvested honey can exceed 19 percent. Yeasts which are present in the honey can activate fermentation. Fermented honey tastes bad. It is fizzy and bitter in taste.

Also do a check for any symptoms of American foulbrood whenever you remove any frames from your hive. American foulbrood spores can be spread through honey.

Granulated honey

All raw liquid honey will granulate. Granulation is when the honey changes texture and takes on a grainy quality. Granulation occurs because the sugar glucose is unstable as a liquid and naturally forms crystals. Granulated honey will often have a layer of watery honey on the top of the jar and thick grainy honey at the bottom. Granulated honey is not spoiled and is still edible. In fact granulated honey is a sure way to know that this honey is raw and has not been overheated or processed so it retains all its natural enzymes.

Early autumn management of your hive

Once a responsible amount of honey has been harvested from your hives it is time to check varroa levels and perform a varroa treatment. Your hive has filled itself with brood and the potential population of varroa is at its maximum. It is important to do a treatment in late summer rather than waiting until late autumn for a number of reasons. By treating early you have the opportunity to re-treat if you find that the treatment you used did not lower the mite population to an acceptable level. This level could be gauged as less than three mites falling over a twenty-four hour period on a sticky floorboard.

The other important reason is that you want a healthy population of worker bees leading into winter. By treating in late summer this gives the hive a full brood cycle of 21 days or longer to raise a worker brood, which are not hampered with viruses spread by varroa, who will lead the hive through the long winter months. Follow the manufacturer's instructions and always remove the treatment from the hive after the set time. Leaving a treatment in a hive will only lead to the worsening problems of mite resistance to certain chemical families.

Tutin toxic honey

Tutin toxic honey is a risk that New Zealand beekeepers need to be aware of, especially if they live in the North Island and the upper region of the South Island. This toxic honey is produced during summer when the passion vine hopper insect feeds off a native shrub called tutu (*Coriaria arborea*). The shrubby native plant is most often found on regenerating bush, roadside cuttings and along streams and in gullies. When the passion vine leaf hopper feeds of this bush it produces honey dew—a sticky sap from the undigested plant material. Bees will collect this honey dew and process it into honey. This toxic honey dew has no effect on bees and the honeydew honey looks, smells and tastes like any other honey. This toxic honey is very stable and can still be very toxic to people who consume it many years after it has been bottled. According to the New Zealand Food Safety Authority around one-quarter of honey samples from the North Island and top of the South island contain some level of tutin.

Be on constant lookout for this tutu plant around your region, if you are a Kiwi beekeeper.

To produce this toxic honey key factors need to be present:
- Large numbers of tutu bushes growing within foraging distance of beehives.
- Large numbers of passion vine hoppers need to be feeding off the tutu bush. This is normally from January over the hot summer months.
- Honey bees need to be present in the area and honey being produced from these hives.

Hot dry summers pose an increased risk as the honey dew is not washed off the leaves by rainfall.

All beekeepers need to manage the risk as the effects of consuming toxic honey include convulsions, vomiting, stupor and coma. In extreme cases death has been documented. There are some ways that you can manage the risk:
1. Take all your honey off before 1 January.
2. Learn to identify the plant and the passion vine hoppers.
3. Do not consume, gift or sell any honey from your hives if it has been harvested after 1 January and has not been tested.
4. Get your honey tested by a laboratory.
5. Become familiar with the legal requirements.

For my total peace of mind I get all my honey tested each year as I find my local nectar flow is much later than December and often the bees only start bringing in honey to store in late January and February.

Getting your honey tested is simple. Once I have crushed and strained my honey I pour around 1¾ fl oz (50 ml) into a small lab specimen jar and label it with the hive name. As I have several hive locations around my area I send these all off to the lab and request a "composite test". This is a cheaper option than having each individual sample tested. Up to ten samples can be tested together and the testing equipment can identify whether one or more of the composited samples may be above the limit for tutin. If a positive result is found further testing is required to identify the individual samples responsible for the high result. Any New Zealand honey sold commercially follows these strict guidelines and is thus always safe to consume.

Late autumn management of your hive

Nectar sources for honey bees in autumn
- Aster
- Flowering gum
- Sedum
- Fennel
- Ivy

Taking care of your colony in the autumn will ensure that your bees will successfully overwinter and come out into spring strong and productive. It is important to ensure the bees going into winter are strong, healthy and disease-free as these bees will live right through winter caring for the queen.

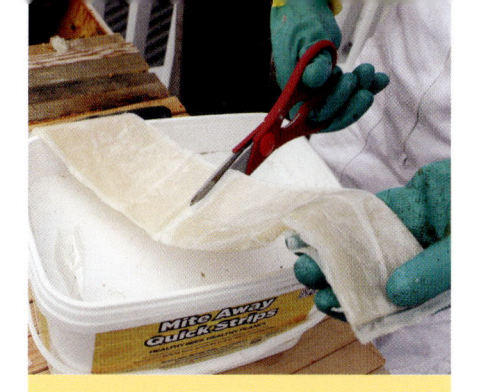

Wearing acid-proof gloves, cut the Mite Away Quick Strips in half and then place them in your hive.

In a Top Bar hive, the Mite Away Quick Strips are placed inside the hive on the floor under the brood area. The hinged bottom floor is closed during the treatment.

During your hive inspections in autumn you need to check:
- The amount of honey that is in the hive for the bees to feed off during winter. You should be aiming to leave the majority of your bees' honey for them over winter in their first year to ensure they do not starve. At a minimum leave at least six full frames of honey for your bees.
If your bees do not have enough, feed them a sugar syrup solution of two parts white sugar to one part water by weight or alternatively feed white granulated sugar crystals inside the hive, directly above the colony (in a Langstroth-style hive) or directly below the brood nest on the floor in a Top Bar hive.
- Are the frames of comb in good order? Remove any combs which are poorly drawn or have been damaged. In a Top Bar hive move any empty drone brood comb and place it at the furthest end from the brood. Come next season's honey flow these combs can be utilized by the colony for honey storage. Remove any frames which have been damaged. Your objective is to ensure the bees can expand next spring on clean worker brood comb.
- The pattern and extent of the brood frames. Is the queen still laying in a uniform pattern without many empty, missed cells? On each frame is there a semicircle of capped honey at the top with an arch of pollen underneath, followed by a good packed area of brood. If there is a lot of worker brood ensure that you are leaving adequate honey for these bees.
- What is the health of the adult bees and brood like? Conduct an American foulbrood check on your colony and treat your colonies for varroa mites so they go into winter strong.

Placing a homemade sticky board made from a real estate sign under the mesh floor allows Sonya to monitor the mite fall during the Mite Away treatment.

Sonya and her Top Bar hive

After harvesting a small amount of honey Sonya was ready to do an autumn treatment for varroa and prepare her hive for the long winter months. It is best practice to get your autumn varroa treatments in as early as you can. In the Southern hemisphere it is recommended to get your treatments underway by the end of February. This allows for a full brood cycle to be adequately treated and to hatch into healthy strong bees and be able to carry the hive through winter in a strong fashion. By treating early you also have the opportunity to re-treat before winter if you find that the first treatment was not effective. This becomes more and more important as varroa develop resistance to some of the mainstream, widely used chemical varroa treatments.

Sonya has chosen to use an organic treatment called Mite Away Quick Strips. The active ingredient is formic acid, in the form of a vapor-releasing strips. This treatment can be conducted whilst honey is in the hive but it is recommended not to harvest any honey for consumption until two weeks after the end of treatment. The quick strips come in pairs enclosed in a plastic sachet. Sonya cuts this carefully with scissors and then carefully removes the strips wearing acid-proof gloves and eye protection. The strips are then separated into two by cutting down the middle section. This formic acid treatment is incorporated into a white colored saccharide gel strip between a paper wrap. When the formic vapors are released it kills both the male and female varroa under the brood cap as well as mature varroa on adult bees. The treatment period is seven days. After this time the bees may dispose of the strip outside the hive or you can remove it and compost them if you wish. The strips do not have to be removed as they become inactive after seven days. This is handy if your hive is somewhere you have to travel to.

As with many organic acid treatments, the daytime temperatures are critical. Do not apply this treatment if the day temperature rises above 85°F (29.5°C), as the vapors will be released too quickly potentially killing emerging brood and the queen. Aim for a temperature during the day of highs between 50–85°F (10–29.5°C).

Sonya placed the strips along the bottom of the inside of her Top Bar hive underneath the brood section of her hive. She ensured that the hive's entrance holes were fully opened but the hinged bottom floor board was closed. Immediately after the strips are placed inside the hive the bees began a roaring noise and they fanned the vapors around the hive. She followed the recommendation of disturbing the hive as little as possible during the seven-day treatment.

Sonya placed a sticky board under the hive floor to monitor the mite fall. After the treatment she will conduct a sugar shake test to monitor the effectiveness of this treatment.

Preparing your hives for winter

Wintertime is the most challenging time for new beekeepers. Beekeepers who have kept colonies between one and two years lost significantly more overwintering colonies than beekeepers who have kept bees for more than two years according to the latest Bee Informed Survey (beeinformed.org/results/the-bee-informed-partnership-national-management-survey-2014-2015). The two key concepts to remember for your bees to survive winter are to leave them ample food stores and manage your bees so they enter the winter months strong in population and disease free.

Overwintering bees are physically different from summer bees. They have more body fat reserves and they live longer as they have little brood to feed and food to forage for. The bees which go into winter in the hive will be the

You can clearly make out where the cluster is located within this Top Bar hive due to the melted frost area on the roof. The honey frames have been positioned to the right of the cluster so they can move to them in a combined unit. By the end of winter the cluster of bees will have moved to the middle or even the far right hand side of this hive.

same bees to emerge and start foraging in the spring. They also form clustering behavior, which is similar to hibernation in mammals.

During cold winter weather the colony of bees will form a cluster around the queen and any remaining brood. The worker bees will rotate so that bees do not spend a long time on the outer margins and risk chilling. Around and in the cluster there are "heater" bees, which vibrate and produce body heat that is distributed through the cluster. The configuration of your hive going into winter is important to ensure that your colony can access food during the cold weather as it is clustering in a tight ball. You need to configure your hive combs so that the cluster can move as a unit to its honey stores. In a Langstroth or Warré hive it may be up directly above their heads. In a Top Bar hive you need to position all your honey combs on one side of the cluster so the cluster of bees can move its way along the food stores over the cold winter months.

During moderate winters bees will leave the hive to forage and hive bees can move honey in storage closer to the cluster. During colder winters the cluster has to become more compact and the entire cluster will move, seeking to maintain constant contact with stored honey. To survive winter a proportion of the cluster must stay in constant contact with honey, as this is the fuel that is needed to provide energy for muscles, which generate heat.

In hives that have died over winter, it is not uncommon to see the dead cluster of bees just inches away from honey stores they have not managed to access. For this reason always remove the queen excluder for winter if your hive is more than one box tall.

It is important to not open your hive unnecessarily over winter. Every time a hive is opened the propolis seal is destroyed, letting in cold air, moisture or pathogens. Observe the health of your winter bees by observing any bees flying during warmer days, looking through the observation glass, if your hive has one, observing the debris on a sticky board under the screen bottom board or by lifting a corner of the hive to check to see that it is still heavy which normally means there is adequate stores of honey.

Many beekeepers go to extreme lengths to prepare the hive for winter, especially if they live in regions that experience extreme weather. These measures include wrapping the hive in insulation material, moving hives to warmer locations or even inside, or placing extra insulation on top of the hive. The Bee Informed National Survey discovered that there was no significant difference in winter colony survival between beekeepers who did conduct extra winter preparations and those who did nothing. (beeinformed.org/results/the-bee-informed-partnership-national-management-survey-2014-2015). Perhaps the single most important thing is to ensure that your bees have ample honey stores in the hive and it is positioned right next to the cluster of bees and your hive has a high population of healthy bees.

CHAPTER 9

PEST MANAGEMENT

Varroa mites

Bee mites have become the bane of all beekeepers around the world. The only main beekeeping country that does not have these mites is Australia, and let's hope it stays that way. The effects the mites have on a bee colony now sees nearly 25 percent of all beekeepers lose more than half their overwintering hives with additional losses throughout the year (Caron and Connor, 2013). Hives infested with mites collect less nectar and are less efficient when it comes to pollination due to their weakened state. Some beekeepers have suggested that even an infestation of 1 percent will reduce the colony's vigor resulting in less honey stored.

Mites are eight-legged brown colored insects. They can just be seen with the naked eye and measure around $1/16$ in (1.6 mm). They look like turtles with a hard, curving exoskeleton. These mites used the Asian bee *Apis Cerana* as a host and this species of bee has developed ways to survive with these mites. They will seek out these mites and physically remove them from other bees, out of cells or from the comb and discard them outside the hive. Unfortunately our European honey bees have not had this evolutionary time to develop any adaptations.

Two varroa mites on a worker bee. If you see this in your hive it normally means that your varroa levels are dangerously high and your hive could crash and die within weeks. Immediate intervention is required. If you see this sort of infestation of varroa on adult bees your hive is on the verge of collapse. This worker bee is exhibiting signs of deformed wing virus(crinkly wings) which is spread by varroa.

This colony has died from varroa. If you look inside the cells you can see white and yellow specks on the inside of the cell wall. This is varroa feces and gives you a clear indication of what your colony died from.

The varroa lives and feeds off the hemolymph (best described as the bee's blood) of the adult bee and the pupae. Perhaps most importantly, the varroa mite spreads bee virus throughout an infected colony.

Mites prefer drone brood to worker brood as the mite in the drone brood has three more days inside the enclosed cell pupating. It has been suggested that this preference for drone brood can range between 3–12 more mites invading drone cells than worker cells (Jandricic and Otis, 2003).

This gives the young developing mites a longer time to feed off the developing pupae. In worker brood a female mite can produce one daughter mite. In drone brood, due to the longer development, time the mother mite can produce up to three mature daughters.

In a very heavy infestation you can see varroa attached to the bodies of

If you have good eyesight you may see live adult varroa mites on your bees. In this image the bee in the middle has an adult varroa just under her left wing. If you don't, never assume your hive does not have varroa. PHOTO: *Alphapix.*

adult bees. If you see this then your hive may be within imminent collapse. This is for the simple reason that around three-quarters of the mite population is actually developing under the cappings so you are only seeing a very small population of them.

The emerging bees who have shared their cell with developing mites emerge with crinkled wings, a shorter abdomen, less hemolymph volume and a much reduced body weight. All these factors will cause this bee to be removed from the hive by other bees or die prematurely. If you observe bees crawling outside the entrance of the hive with misshapen wings you have a varroa problem in your hive.

If you don't see varroa mites in your hive, don't think that you don't have them. Every hive in every country that has varroa will have mites. If you keep bees in your garden you also keep mites too. Nowadays we need to recalibrate our ideas about mites. We will never have hives that are totally free of mites; rather, we, and our bees, need to learn to live with a level of mites that is non-lethal to the bees.

How to monitor for varroa levels within your hive

As backyard beekeepers we are lucky in that we have the time to monitor closely for building mite levels. These are the ways I monitor my hives for mite levels:

A sticky board: All my hives have a mesh bottom floor and a hinged bottom board that I can close. The stainless steel mesh is small enough that bees cannot get through it but large enough that any mites which fall off bees will fall straight through. I trim an old Corflute real estate sign to size, spray it with some cooking oil and insert it between the mesh floor and the solid floor board. The idea behind this is that any mites that fall off bees will fall through onto the white sticky board and be stuck there. I return in 24 hours and count how many mites have fallen. This will give me a natural mite fall. A twenty-four hour natural mite fall of eight to ten is considered by most to be the threshold for treatment (Conrad, Ross). Danish research suggests multiplying the daily total mite fall by 120 to reach the total varroa population in the colony during the production season of summer to early autumn (Goodwin and Taylor, 2007). The level where varroa start to take a real toll on the health of the colony is thought to be around 3,500 mites per hive at the end of summer.

Uncapping drone brood: As most varroa live under the drone brood cappings, you can monitor their numbers by uncapping some drone brood during your hive inspections. Randy Oliver sates that varroa prefer drone brood to worker brood by a ratio of about 10:1 (scientificbeekeeping.com/fighting-varroa-reconnaissance-mite-sampling). You can buy specially designed uncapping forks from the bee supply companies for this job but I just use my hive tool or kitchen fork. Choose some of the older capped drone brood. Older brood cappings will be darkened. Gently flick the caps off with your fork and pull out the pupae. Observe the pupae. Can you see mites moving over the white body of the pupae? Observe the open cell—often you will observe mites crawling out of the now open cell. If most of the drone pupae you remove have multiple mites, a treatment is called for.

Sugar Shake: This method of measurement is non-invasive and will not harm any bees. The theory is to cover a number of bees in fine sugar, which makes them very slippery. Any varroa will be dislodged and then these varroa can be counted to give an indication of how many mites are present in the hive.

Equipment required
    1 wide-mouthed preserving jar with a screw ring
    A circle of wind (hardware) cloth with holes wide enough to allow the passage of mites but not the bees
    2 tablespoons of confectioners' (icing) sugar
    A white surface such as a dinner plate
    A spray bottle filled with water
    A 4 fl oz (125 ml/½ cup) kitchen measuring cup (this measures approximately 300 bees)
    A black permanent marker

Method
1. Fill the 4 fl oz (125 ml/½ cup) measuring cup with water, tip into glass jar and measure level with permanent marker on outside of jar. Collect some adult bees and quickly pour them into the wide-mouthed jar and contain with mesh and a screw ring. Collect the bees from at least three different frames within the brood area and make sure you don't collect the queen.
2. Add 2 tablespoons of confectioners' (icing) sugar into the jar.
3. Gently roll the jar for 4 minutes so that all the bees are covered in the fine sugar. Do this in the sunshine as the sun will increase the humidity inside the jar and this will help dislodge the mites.
4. Rest the jar in the shade for 2 minutes (this gives the bees a chance to further dislodge any mites).
5. Upend the jar over a white surface and firmly shake out the powdered sugar. You need to be firm and shake hard to ensure all the mites are dislodged.
6. Spray the plate with the water spray to dissolve the sugar so the mites are visible.
7. Count and record the number of mites.
8. Return the jar of angry but unharmed bees to the hive. They will clean themselves.

What the results will tell you

To determine the percentage infestation level, divide the number of mites by the number of bees in the sample. So if you find twelve mites in the jar (filled with 300 bees) the equation will be 12 ÷ 300 = 0.04 which is 4 percent.

As a general goal try to keep the mite level lower than 2 percent infestation in adult bees and never allow it to climb to higher than 5 percent.

Using integrated pest management to control mites

Integrated Pest Management (IPM) is a multi-pronged approach to pest control. IPM uses several non-chemical strategies to manage a pest, in this case varroa. It is not realistic to think we can totally eliminate varroa in all our hives, so by devising ways to keep numbers at a non-lethal level our bees may learn to live with them. The following are some integrated pest management interventions that you could incorporate into your own hives.

This white board which was placed under the mesh-bottom board of a hive has been sprayed with some cooking oil to make it sticky. After 24 hours it can be removed and a varroa count can be done. The small, oval, brown shapes are varroa which have fallen through the steel mesh and have been caught on the sticky board.

Mesh-bottom board: A mesh-bottom board permits any mites which have fallen off bees or been removed by the bees to fall through the mesh out of the hive. Without a mesh-bottom board the mites can fall to the solid floor and then just jump on another passing bee. Mites can jump 1½ in (40 mm) so it is important to have the hive base raised more than this off the ground. To conserve winter warmth I also incorporate a hinged, solid-bottom board in my Top Bar hives that I can close when I choose. In a recent Bee Informed Partnership National Management Survey carried out with American beekeepers, it was reported that beekeepers who used a screen-bottom board lost 10.7 percent fewer hives than those who did not use a screen-bottom board.

Drone brood culling: Research suggests that by removing capped drone brood comb every two weeks throughout summer you can keep mite levels below the dangerous threshold when varroa mites start to take a toll on the hive's health. Of course this action can have a detrimental effect on the number of drones available to mate with queens and the morale of a hive.

As varroa prefer to breed in drone brood you can combine specialized drone brood sized foundation within some of your frames. These larger-sized foundations allows the queen to lay only drone brood in this one specialized frame. When this frame is capped you can remove the entire frame and kill the brood along with all those developing mites. You can do this by placing in a freezer for at least 24 hours. The frame can then be returned to the hive where the hive bees will remove the frozen and dead pupae and varroa mites and prepare the cells for the queen to lay again.

I find my Top Bar frames get too brittle in the freezer and easily break so I place these naturally drawn brood combs in a small, Top Bar hive nuc box. I set this in a hot, sunny spot and pour 2 fl oz (60 ml) of 65 percent formic acid on an absorbent pad before closing the hive for 24 hours. The formic acid fumes penetrate the cell capping and kill the varroa mites. This frame can then be returned to the hive. The bees will remove the chilled and dead brood and polish the inside of the cells in readiness for the queen to relay eggs.

However, this management procedure could be rebuked as there was no significant hive loss difference between beekeepers who employed this practice of drone brood removal and those who did not, according to the Bee Informed Partnership National Management Survey 2014–2015 in the United States (beeinformed.org/results/the-bee-informed-partnership-national-management-survey-2014-2015).

Apiary sites: Colonies with more sun and good air drainage have fewer varroa mites and other pests such as small hive beetles (Caron and Connor, 2013). Place your hive facing the sun and on the upper slope if your garden is steep. Provide protection from the prevailing wind but away from spreading deciduous trees, which can throw too much shade.

Hygiene: Avoid the routine transfer of frames between your hives. If you have more than one hive, position the hive entrances so that they face different orientations. This helps to avoid bees returning to the wrong beehive after a day of foraging. Avoid having more than four beehives in close proximity to each other for the same reason. In the wild, bees choose new hive locations that are a distance from the original colony.

A break in the brood cycle: This is a useful technique that stops the colony from rearing brood and at the same time prevents the varroa mites from reproducing in brood. A brood break can be done by fencing in the queen on a brood frame in an enclosed cage. She can still move around and the hive bees can feed and groom her through the mesh but she cannot travel around the brood nest laying eggs at will. Another way to create a break in the brood production is to make numerous splits during the warmer months. The weeks while the colony is waiting for a new queen to develop allow them time to groom off any mites and prevent more from breeding inside capped brood.

Coordinate varroa control with nearby beekeepers: Varroa levels can build up quickly after a treatment, if surrounding hives are not being treated. A beekeeper may treat their hives in spring and have completed the treatment before a neighboring beekeeper has even commenced. Varroa spread on drone bees, which are more likely to fly to other hives, and also by hitching a ride back on foraging bees. If you know of other beekeepers in your area try and coordinate any treatments so they all coincide.

Treatment-free beekeeping

A growing movement among some beekeepers is treatment-free beekeeping. As the title describes, this is beekeeping without using any sort of treatments. Solomon Parker from www.parkerfarms.biz has been a treatment-free beekeeper for a number of years and runs an informative Facebook page and regular podcast. He defines treatment as "anything done in the hive, introduced by the beekeeper into the hive with the intent of killing, repelling, or inhibiting a pest or disease afflicting the bees, or in any way 'helping' the bees to survive when they ought to be surviving by themselves".

The theory treatment-free beekeepers follow is one of Darwinism. This theory developed by the great English naturalist Charles Darwin, follows the idea that all species arise and develop through natural selection of small, inherited variations that increase the individual's ability to compete, survive and reproduce.

Allowing the bees to naturally adapt to varroa really does make sense. This is what every living thing in the natural world has been doing for millennia. How we keep bees today, propping them up with man-made chemicals, is not sustainable in the long term. The unfortunate truth is that commercial beekeeping could not afford to suddenly go treatment-free, nor could the consuming public. Becoming treatment-free would mean a worldwide collapse of commercial operations.

It really is up to small-scale backyard beekeepers to make the sacrifice and explore these more natural and sustainable beekeeping methods. Perhaps they will be able to build up bee genetics for all of us less brave beekeepers to appreciate.

Organic treatments for your hive

Preventing toxic beeswax

Beeswax is an extremely complex material containing over 300 different substances. It consists mainly of esters of higher fatty acids and alcohols. Unfortunately, within the hive, wax acts like a sponge soaking up contaminants introduced into the hive. Scientists at the American Chemical Society have studied beeswax and identified 87 pesticides found in wax samples (www.sciencenews.org/blog/science-public/bees-face-unprecedented-pesticide-exposures-home-and-afield). Most pesticides are lipid soluble, which means that the bee's wax soaks up the chemicals and stores it, often for decades, in the wax. These pesticides most commonly are introduced by the beekeeper in the form of varroa mite chemical treatments. Foraging bees were found to bring organophosphate insecticide, fungicides, and herbicides back to the hive on their bodies.

Chemical levels quickly build up after each varroa treatment and in a very short time, often only a matter of a few years, sub-lethal levels are found in the wax. A sub-lethal level means that the quantity of harmful chemical will not kill the bee outright but after a time of exposure will cause death. The sub-lethal effects of this toxic cocktail of residues in a hive causes delayed larval development, delayed adult emergence and a shortened life span of the adult bee and queen. These facts are very good if you are a varroa mite though! A longer larvae development time means mites have more time to breed in the capped cell. Being surrounded by wax, which contains levels of the chemicals beekeepers use to try to kill mites, allows the mites to quickly build up a resistance to these treatments. It's a win-win for the mites!

When you consider that the wax comb is the heart of the hive and where eggs are laid, brood is reared and honey and pollen is stored, toxic wax will have a catastrophic impact on any colony.

If you choose to use the conventional varroa chemical treatments in your hive it is recommended to make sure that you always remove the last treatment, only treat when it is necessary and rotate out old frames of beeswax on a regular basis, (at least every 2–3 years) and replace with new frames. As foundation sheets are manufactured from recycled wax I personally recommend not using these in your hives. Allow the bees to build their own naturally formed comb. Researchers at the University of Pennsylvania have found that commercially available beeswax foundation they tested were contaminated with a total of 25 pesticides (mysare.sare.org/mySARE/ProjectReport.aspx?do=viewRept&pn=FS12-261&y=2013&t=1).

The best strategy to improve wax purity and in turn hive health is to only use non-toxic natural organic acids as an alternative varroa control. Oxalic acid and formic acid are naturally occurring acids already found in the hive in minute levels. As they are not fat-soluble, neither are stored by the wax so they do not build up to sub-lethal levels.

If you plan to use your wax from your hives to make skin care lotions or other items to be used with food, ensure that it is from an organically managed hive or, even better, treatment free.

I am not treatment-free but I have made the decision to treat for varroa only using organic treatments. It was only a few years ago that many established beekeepers were telling all new beekeepers that organic treatments simply did not work. Now we are seeing resistance by varroa to some of the conventional chemical treatments. The selection of chemical treatment options is growing smaller and smaller.

Organic treatments use naturally occurring acids that are already found in the hive, albeit used in a higher concentration. Other organic treatments use natural essential oils such as eucalyptus, thymol, and camphor.

When you commit to using organic treatments there are a few things you have to accept. Firstly you will have to treat more often—four or five times a year compared to the twice-yearly treatments using conventional treatments. With organic treatments you need to accept that you will never get a 100 percent knockdown. Organic treatments are more about keeping varroa levels below the threshold where they cause damage to the hive.

Organic treatments are not as convenient as conventional treatments. They require more visits to the hive and they can be very temperature dependent. Some, such as formic acid, require safety equipment and very careful handling.

Organic treatments

These are some of the organic treatments I have used in my hives:

Oxalic acid

Oxalic acid is a naturally occurring acid and is produced in very small amounts in rhubarb leaves. The use of oxalic acid to kill varroa has been used in Europe successfully for many decades. Oxalic acid will not penetrate into the sealed brood so will not kill developing varroa under cappings. Because of this, oxalic acid is best applied when a colony is broodless or several applications need to be repeated to catch a full brood cycle. Oxalic acid will kill varroa on adult bees and it is believed that the hive bees will deposit some of this acid into the brood cells between cycles as they go about their duties, cleaning and preparing empty cells for the queen to lay in.

Oxalic acid is very corrosive so always wear chemical-proof gloves, protective glasses and a gas mask. Store the crystals in a childproof and well-labeled jar.

In areas which experience cold winters, and a period of broodlessness, a 1:1 sugar syrup mixture of oxalic acid and sugar can be carefully syringed over the clustered colony. This is known as the dribble method. The bees consume the oxalic acid and sugar syrup and the varroa then feed off the bees and die.

To make the solution to dribble over the bees use some electronic kitchen scales for accurate measurements. Mix 1½ oz (45 g) of oxalic acid crystals with 1 lb 5 oz (600 g) of white sugar with 21 fl oz (600 ml) of water. This will make a 35 fl oz (1 liter) mixture. Discard any that is not used. The dribble procedure is best carried out when the weather is cold and dry so that the bees are clustered.

1. Before treatment, conduct a natural mite fall so you can compare counts before and after treatment.
2. Fill a 1¾ fl oz (50 ml) syringe with warmed (around 95°F/35°C) oxalic acid solution.

3. Open the hive to expose the clustered bees.
4. Trickle 1 teaspoon (5 ml) of solution between and along each bar of the hive.
5. Close the hive, and monitor mite fall.

In areas that experience a temperate climate, your bees may not experience a broodless time over winter so the vaporizer method will be more appropriate. This is the method I use as I find my bees are active and the queen will continue to lay a small amount of brood throughout winter. This method uses a small element to heat the oxalic acid crystals, turning them into vapor. This vapor spreads through the hive and coats all the bees with a fine mist. It does not harm the adult bees but will kill all the varroa living on the adult bees. Using a vaporizer is also a good method to use when you catch a swarm. You can give a newly settled swarm one application and can be assured that all the varroa (99 percent) will be killed as all mites will be on adult bees.

Recent studies are suggesting that applying oxalic acid via the vaporizer is more effective and less harmful than the dribble method. I find it a very useful method to use mid-season if you are concerned that mite levels are too high in a particular colony. It can be used any time and is not temperature dependent as some of the other organic treatments are. It is best to conduct treatments when there are not honey frames in the hive as it can cause some noticeable tastes in the honey. If you do need to conduct a treatment when honey stores are present, wait for at least three weeks before harvesting the honey or remove these frames whilst conducting a treatment. Oxalic acid has a very short half-life, this means it breaks down quickly so will not build up in the wax. It is naturally found in colonies in very small amounts.

To administer oxalic acid via a vaporizer, you need to purchase a small hand-held vaporizer from any of the online bee stores and a small 12-volt rechargeable battery. You also will need a chemical vapor mask, some chemical-proof gloves and a small measuring spoon.

How to administer oxalic acid via a vaporizer
1. Conduct a natural mite fall.
2. Block up any ventilation or entrance holes in the hive with a damp towel or similar.
3. Wearing vapor mask and gloves, place ¼ teaspoon (packed) of oxalic acid in the vaporizer (per brood box), insert it so it is directly under the colony.

4. Attach clips to a 12-volt battery and leave for two minutes. The vaporizer heats the acid and turns it into a smoky vapor.
5. After two minutes, disconnect from the battery and leave in place for a further ten minutes.
6. Remove cloths and vaporizer.
7. Four to five treatments recommended, every five days for four weeks to cover gestation period of mites. Monitor mite fall.

Treating your hive with oxalic acid

To treat with oxalic acid and a vaporizer in a Warré or Langstroth/Flow hive is very simple. Just insert the element with the acid crystals gently into the entrance at the base of the hive. Stop up any gaps with a damp towel. The hive design will act like a chimney, forcing the vapor up through the colony.

In a Top Bar hive with a hinged-bottom floor, insert the vaporizer between the hinged-bottom board and mesh floor under the brood area and then close the floor and block any gaps with a wet towel. If your Top Bar hive design has a solid wooden floor you may need to retro-fit the hive by cutting a small opening in the end near the brood so you can slide the vaporizer in during treatments. It will require some sort of closing when not in use to keep the hive warm and defensible for the bees.

Formic acid

Formic acid is another naturally occurring acid. Ants produce formic acid in very small amounts. Formic acid is more hazardous than oxalic acid. It is very caustic and will cause deep blisters if it comes in contact with your skin. The alarming thing is you often will not feel it burning but will notice blistering skin after a few hours. Particular care is required when handling this liquid. Wear chemical gloves, goggles and a facemask. It can be lethal if swallowed and the fumes can ignite. The acid smells of vinegar. Formic acid will also corrode metal so do not let it come into contact with any steel hive lids if you use these.

Formic acid will penetrate under the brood cappings and kill the developing mites. I apply formic acid in one of two ways:

Using 65 percent formic acid applied onto an absorbent pad, I pour 1¼ fl oz (40 ml) of formic acid and place it on the hinged-bottom board of my Top Bar hives and close the base. It is important to leave all the entrances fully open. Formic acid is most effective when temperatures are between 50–79°F (10–26°C). In a Warré hive or Langstroth/Flow hive you can apply this absorbent pad on top of the hive top bars. The formic acid vapor is heavy so it will naturally drift down through the hive. The bees' fanning will disperse the vapor to all areas of the hive. This treatment can be repeated up to six times at five-day intervals. Repetition of at least four times is recommended.

This treatment is very temperature dependent and it is good practice to monitor weather forecasts. If a heat wave is forecast, delay the treatments. If the temperature rises above 86°F (30°C) the vapor released will be too high for the bees to tolerate. Remove any treatments from colonies during a heat wave to prevent excessive brood mortality, risk of killing the queen and absconding.

Sometimes you will observe some emerging brood loss. A newly inserted treatment may kill brood as it is just emerging from the cells. This is especially evident in a single brood colony. Treated colonies may have temporary suppression of population growth, which the colony will recover from. Do not use formic acid on very small colonies of less than five frames as this could result in excessive colony damage.

When working with formic acid ensure you do this in an open area, stand up-wind and avoid breathing the vapors. I always ensure I have some fresh water on hand at all times to wash any accidental spills. Store formic acid in its original container away from sulfuric acid, sparks or flames.

Mite Away Quick Strips

This formic acid based treatment comes as an absorbent pad impregnated with formic acid. The acid is packaged in a unique bio strip with the formic acid incorporated with a sugar-based material. When placed in the hive, the formic acid is released and the bees eat the sugar and dispose of the material out of the hive. This eliminates the need to re-enter the hive to remove the treatment as the bees do this. This product also is permitted for use whilst honey supers are in the hive. This treatment will target the mites under the cell cappings.

It is important to wear chemical-proof gloves when handling this product, as it is very easy to suffer blisters and burning to your hands if you touch this product unprotected.

Mite Away Quick Strips being applied to a Top Bar hive in early autumn whilst the days are still warm.

A high mite fall is evident on this sticky board after treatment using formic acid.

Thymol

Thymol is an essential oil extracted from the thyme herb plant. This essential oil is widely used to kill varroa. It is very effective at killing mites and not harming the bees. Again the vapors of this treatment do not penetrate the brood cappings so the treatment needs to be in place over a full brood cycle of 24 days. The external air temperature needs to be above 59°F (15°C).

It is best not to treat when honey is in the hive as the thymol vapors will cause noticeable taste residues in honey and wax. However these residues do not persist for long. It is recommended to treat all colonies in your apiary at the same time as thymol treatments can cause robbing. During treatments I find that the bees can become more aggressive and when a fresh treatment is added to a hive, the bees will fan loudly and will often hang on the outside of the entrance to escape the vapor.

I use Thymovar wafers. These yellow wafers can be cut and inserted between the top bars of my Top Bar hive. I position them so they are near the sloping sides of the hive. This helps to prevent emerging brood being killed by having the wafer directly on top of them. I close the bottom floor board of my Top Bar hive when a treatment is in place. One wafer cut in half is placed in the hive when temperatures are between 54–86°F (12–30°C). Its effectiveness is reported to be between 82–100 percent. A repeat treatment is carried out in a further four weeks' time. I have been impressed with its efficiency.

Thymol is an organic treatment which is easy to use but is temperature dependent. It will not work effectively in cold weather.

To apply a thymol treatment in a Top Bar hive, cut the stripe in half lengthwise and insert between the top bars in the brood area of the hive. Repeat this treatment in four weeks' time.

ApiLife Var

This organic treatment is a combination of thymol, eucalyptus, camphor and menthol impregnated into florist foam. This tablet is crumbly so when using in a Top Bar hive I insert it into a plastic gauze bag and insert pieces at either end of the brood frames. In a Warré or Langstroth/Flow hive you can place these tablets on the top bars. A further treatment is recommended after 3–4 weeks. Daytime temperatures should not fall below 59°F (15°C) for a long time. Take off any harvestable honey prior to treatments so as not to taint the honey with a strong eucalyptus taste.

Apiguard

This organic treatment comes as a gel formation of thymol. It is designed to give a controlled release of vapors. Bees will pick up the gel and move it through the hive. Some overseas trials have showed a 98 percent effectiveness. The volatility of the product is temperature dependent and requires temperatures between 59 and 68°F (15–20°C). In a Top Bar hive I open the foil pack and place this on the hive floor near the brood cluster. I keep the hive floor closed during treatment. With a Warré hive or Langstroth/Flow hive, you can place this opened foil packet of gel on the top bars where the bees will access it and spread it around the hive.

Some of the organic treatment options

NAME	ACTIVE INGREDIENTS	TIME OF YEAR & TEMPERATURE PARAMETERS	DURATION OF TREATMENT	NOTES ON USE
Thymovar ®	Thymol	Use late spring and early autumn when daytime temperatures are between 54–86° F (12–30°C).	1 wafer cut in two placed on top of frames of a Warré, Lang or Flow hive per brood box. Insert into the brood area in a Top Bar hive. Repeat the treatment in three to four weeks.	– Treat all hives at the same time as can cause robbing. – Keep unused wafers in air tight sealed packet. – Can cause the bees to become more aggressive during treatment. – Keep entrances open but close bottom floor board on your Top Bar hive.
ApiLife Var©	Combination of thymol, eucalyptus, camphor and menthol infused in a vermiculite wafer.	Daytime temperatures should not fall below 59°F (15°C) for any length of time. – Treat at the end of summer when honey has been removed.	Use one tablet cut into four pieces and place on the bars of each brood box. In a Top Bar hive I place these pieces into small gauze envelopes and insert into the hive surrounding the brood area. Repeat this treatment after three to four weeks.	Keep entrances open but close all mesh bottom boards.

Formic Acid	Formic Acid 65 percent solution flash treatment	When temperatures are between 50–70°F (10–26°C).	– 1¼ fl oz (40 ml) of 65 percent formic acid poured on to an absorbent material and placed in hive. Repeat this every five days for at least four times.	Close mesh floor but always leave entrances open. If temperature rises above 79°F (26°C) remove treatments from hive.
Miteaway Quick Strips	Formic Acid	Outside daytime temperature highs should be between 50–85°F (10–29.5°C) on day.	Treatment period is seven days.	Hot temperatures – above 91°F (33°C) – during the first three days may cause excessive brood mortality and queen loss. Some brood mortality and bee mortality may be observed. Do not disturb colony during treatment.
Oxalic Acid via dribble method	1½/2 oz (45 g) of Oxalic Acid crystals 99 percent mixed with 1 lb 5 oz (600 g) of sugar and 21 fl oz (600 ml) of warm water.	Conduct dribble method in winter when there is no brood and the bees have formed a cluster.	Dribble 1 teaspoon (5 ml) of the warmed solution along and between each bar of the cluster.	Use chemical proof gloves. This method can only be used when a colony is broodless.
Oxalic Acid via vaporizer	Oxalic acid crystals 99 percent	Not temperature dependent. Will not kill varroa under capped cells.	Place ¼ teaspoon of oxalic crystals onto the vaporizer plate insert into base of hive and heat for 2 minutes. Disconnect from battery but leave hive closed for a further ten minutes.	Stop up any gaps. Repeat treatment every five days for at least four weeks to catch the full brood cycle. A good choice of treatment if a hive requires a mid-season treatment.

The bees on this Flow frame have built out all the gaps on the plastic frame. The bees are now filling each cell with foraged nectar. When the cells are full, excess water will be evaporated and the cell sealed with a wax capping.

Sonya and her Top Bar hive

After Sonya removed the spent Mite Away Quick Strips she conducted a sugar shake to monitor the varroa levels. She was happy with her results so preceded to winter down her Top Bar hive. She ensured that all the brood frames were together and the six combs of honey stores where positioned right next to the cluster on one side so the bees could move together in one direction over the winter months as they consumed the honey stores. As Sonya's hive is on a high balcony she was concerned about chilling winter winds. She decided to provide some extra insulation for her bees. Between the hinged bottom board and the mesh floor she placed a strip of bubble wrap plastic to help insulate. On top of the top bars and under the roof she placed newspaper sheets, approximately 20 sheets thick, to form an insulating layer before the roof was placed in position.

Sonya stopped up one of her three entry holes with a wine cork to help with cold air exchange. She then left the bees alone during the coldest winter months.

This is the equipment required to treat a hive using an oxalic acid vaporizer. You require acid vapor-proof respirator mask, oxalic acid, a small portable 12-volt battery, a vaporizer and some spray on cooking oil to make a sticky board.

Sarah and her Flow hive

Sarah's bees had started to wax up the Flow hive frames and placed a little nectar in some of the lower cells but, as winter closed in, the nectar flow from plants around her neighborhood ceased.

Sarah has noticed a few varroa dropping onto the sticky board under her hive and has chosen to treat using oxalic acid via a vaporizer. She has chosen this treatment due to it being a softer, organic option and the fact that the oxalic acid has a very short half-life and does not stay in the wax.

Sarah will do four applications every five days and then assess her varroa counts using a sticky board under her hive and counting the varroa numbers after twenty-four hours.

As the vaporizer could not fit into the front entrance, Sarah treated the hive from the rear under the mesh floor.

An easy way to make a sticky board for varroa monitoring is to cut a white real estate Corflute sign to size and spray it with cheap cooking oil. It can be wiped and reused many times.

Sarah with her protective equipment carrying out an oxalic acid treatment.

Eric and his Warré hive

To winter down his hives, Eric has gone through all the boxes and condensed the brood frames and the queen in the lower two boxes. He has placed a box of capped honey on top of these. He has removed the queen excluder so that the queen can travel up into this box if she needs to. Eric is not sure how much honey his hives will require over winter so has some extracted honey from his hives which he will use to feed back to his bees if required.

As part of his Warré hive design, the entrances are small and as he has not seen any wasps or robber bees, he does not see it as necessary to close them down any further. Eric does not need to place insulation under the roof as the sawdust-filled quilt box already does this.

Eric is using oxalic acid via a vaporizer for his varroa control going into winter.

A mouse has taken up residence on top of this Top Bar hive during winter.

Mice will get the comb to get to the pollen and stored honey.

Pests and diseases of the honey bee hive

Pests

Mice: These rodents can be a problem during winter when they seek out a warm, protected place to nest. During winter the bees are in a tight cluster so cannot defend all their combs. Mice particularly love the curving pollen band at the top of the comb. They can make a mess by hauling nesting material into the hive, leaving their droppings and chewing on the wax combs. A mouse can squeeze through any hole as wide as a mouse's skull. I have found a mouse nest on top of the top bars under the roof of my hive.

If you observe your screen bottom board and it has chewed bits of wax, small pieces of grass, leaf matter or small brown pellets (mouse droppings), a mouse may be nesting inside your hive. Further investigation would be warranted. With hives that have the entrance on the ground, reduce the size of the entrance to less than 1.2 in (3 cm) to prevent the rodents from entering during winter. You could lay some baits under the hives, protected from mammals, in approved traps. A safer, non-chemical

This telltale webbing is a sure sign you have wax moths in your hive.

This comb has been destroyed by wax moths. Remove from the hive and feed to chickens if you have them or freeze for at least 24 hours.

technique is to attach a mousetrap. Cut some small gauge wire netting and use some flathead screws to cover the entrances of the hive. Remove in spring.

Cockroaches and spiders: In some climates these insect pests are just something that comes with a hive. When I open up my Top Bar hives it is common to find cockroaches living on top of the top bars, enjoying the warmth of the hive. They seem to rarely venture inside the hive as perhaps the bees chase or sting them. They do not seem to do any harm.

Wasps: German and the common wasp (called yellow jackets in the USA) are the two species that cause problems with bee hives. They can be identified by their smooth, yellow and black striped bodies. They fly in a fast zigzaggy pattern. They will make their colonies often in a bank or hole in the ground. Wasps are a particular problem in autumn as their nests at this time are at their strongest and bee colonies are starting to decrease in population in readiness for winter.

Wasps do not fly a great distance from their nests to forage, so if you have a large number of wasps around your hives or bothering you when dining outside it pays to try and find their colony. Try to follow their flight path back. Often it is easier to do this at dusk when the sun is low in the sky and they are backlit.

If you locate a nest, observe caution as they can be very aggressive and can sting you numerous times. Wait until after dusk and all the wasps have returned and stopped flying. Wearing your full bee suit and gloves, pour poison into the entrance of the nest and cover. I use an insecticide spray, which I mix up in a garden sprayer, and push the nozzle into the hole if the nest is in or around a building. If the wasp colony is in the ground, I pour a cup of petrol into the hole and quickly cover with a piece of wood or rock.

If wasps are continuing to bombard your hive, try placing a robbing guard in the front of your hive. Doing nothing can mean that your hive will quickly succumb to robbers and you could lose your bees.

Ants: These can be a real problem in some areas. Often I open my Top Bar hive lid and discover a large ants' nest living on top of the top bars. Ants will try and rob honey from the bees and they can taint the honey with their telltale formic acid odor. Try sprinkling powdered cinnamon on top of the bars of your Top Bar hive or on the crown board of your Lang/Flow hive. Some beekeepers have designed their hives with metal legs that are then placed in cans filled with oil. This forms a moat, which stops the ants from accessing the hive.

Wax Moths: There are two species of wax moths which cause damage to beehives. The Greater Wax moth (*Galleria mellonella*) and the lesser wax moth (*Achroia grisella*). The wax moth thrives in warmer regions of the world. It is the caterpillar of the moth's lifecycle which causes the most damage in the hive. These caterpillars feed on the bee larvae cocoons in brood comb, pollen and honey. As the caterpillars move around the frames they leave a web trail, quickly destroying this important resource of drawn comb. Comb which has reared brood or has stores of pollen is the most attractive to these pests. The wax moth is a particular pest for Top Bar hive keepers as the wax caterpillars will tunnel through the pollen and brood area of the comb weakening it and cause the comb to come away from the top bar.

Wax moth are not a major problem in a strong hive but in a hive which is battling another disease or has an ailing queen and thus is low in population, it can quickly become infested with caterpillars, adult moths and comb full of silky webs and fecal debris. When the caterpillars are fully-grown they will pupate into larvae and spin a cocoon. They will find corners of the hive and eat into the wood. It you see gnawed depressions in the wood, these have been caused by wax moth cocoons. A high infestation can weaken your hive woodenware.

The best defense against wax moth is to maintain a strong, healthy hive. Limit the amount of drawn comb in a hive to what the bees can patrol. In warm weather an empty hive full of comb can be destroyed in as little as ten days. If a colony dies, restock the hive as soon as you can with more bees after identifying that the bees did not die from American foulbrood.

Moths need a warm, protected dark space to multiply. To protect any drawn brood comb from the wax moth, first place in your freezer for at least 24 hours to kill any larvae or eggs and then store in a light, airy space such as hanging under a carport until required.

If you enjoy comb honey from your hive, freeze these portions until required to kill any wax moth eggs which may be active in the comb.

Diseases of the hive

No bee diseases will harm humans or render any product from the hive unfit for human consumption but some diseases will mean your bees will die.

Every time you open your hive and look at a frame containing brood you should automatically check for diseases. Within a short time this needs to become an automatic habit for all backyard beekeepers. To proficiently check you need to shake or sweep the majority of the bees off the face of the comb so you can observe the brood. If your eyes are failing, have a magnifying glass or your reading glasses nearby. Hold the frame up and have the sun at your back to help make your assessment easier.

Have this checklist in your mind every time you observe a frame of brood:
- Look at the cappings. Is the color and shape uniform across the frame?
- Does the pattern of the brood create a uniform and dense arrangement of similar aged brood?
- Does the brood appear very spotty with lots of empty cells between capped cells?

- Is the brood capping slightly convex with no holes in the surface of the cap?
- Is there any odor present?
- Are larvae white, lying in the center of their cells and glistening in the light?
- Is the larvae you see glistening white, plump and have the typical segmented body sections?
- If you pick the capping layer off a cell and pull out the developing larvae what color is it? Does it come out in one portion?

When you become confident at identifying healthy brood then any change will be picked up sooner rather than later by your expert eyes. Off-color larvae are the first signs that something could be wrong in your colony and should draw your attention to investigate further.

The following is not an exhaustive list of diseases of the honey bee, but they some of the most common. In your first year of beekeeping you will be very unlucky to encounter any one of these diseases but it could happen. If you are at all concerned about something then contact a more experienced beekeeper to offer a second opinion and advice.

American foulbrood

Let's start with the big guns first! American foulbrood can strike fear into many a beekeeper, be they weekend backyarders or large-scale commercials. In New Zealand any hive which is found to have American foulbrood is required by law to have the bees killed and the hive destroyed (or treated in an expensive and specialized way).

American foulbrood is caused by a bacteria called Paenibacillus larvae. It is a very resistant and long-lived spore. The spores are resistant to boiled water, sun and chemicals and it is believed that they can stay viable in the soil for up to 50 years.

The cycle of this disease commences when a bee larvae eats some Paenibacillus larvae with its food. The spores multiply in the gut of the larvae. The larvae will die at the larval or pupal stage. The younger the larvae the more susceptible they are to this bacteria. The decomposing larvae can contain 2,500 million spores so this disease is quickly spread around the colony (Matheson and Reid, 2011).

Any bee colony is at risk of AFB. Strong colonies perhaps more so, as they are more likely to rob out hives that are weakened by this disease.

The newly updated website of the New Zealand Management Agency www.afb.org.nz has informative videos and a mock test you can take online to test your understanding of this disease.

The signs of American foulbrood

Holes in cappings: Small holes with jagged edges can mean that the hive bees have chewed the cap of infected larvae.

Sunken cells: Infected larvae which have died from AFB are often sunken and have a greasy sheen to them. They are often darker in color.

Color of infected larvae: Infected larvae or pupa droop down into the lower portion of the cell. They change from white to yellow to a coffee-brown color that is consistent over the whole length of the larvae. The larvae lose their segmented look.

Odor: When the disease is advanced the colony emits a fishy odor, which is very distinctive.

Brood pattern: As larvae die the brood takes on a very spotty pattern.

Tongue: If the brood dies in the pupal stage its tongue will point upwards towards the top of the cell wall. After all the remains have dried, this tongue will be evident. This is known as scale.

Roping out: As the larvae or pupa die, the remains become very sticky, almost like toffee. If you poke a matchstick into the cell and swirl it around and then slowly withdraw it, the diseased brood will come out attached to the matchstick but still in the cell as a long brown thread. The brown thread is like elastic and will contract back into the cell if it breaks off the matchstick. Dispose of the matchstick covered in infected material in a hot smoker.

As a beekeeper I have seen American foulbrood once in my beehive. It is a very devastating thing to have to deal with. I can understand how easy it could be for a beekeeper to fear the worst but not act, hoping if they just shut up the hive it may go away. Perhaps if you transfer the brood into a new box or shake all your bees off the frames into a new hive this will resolve the dilemma? All of these actions will do nothing but continue to spread this disease through your remaining hives or to other nearby apiaries and often these actions are illegal. Act quickly and responsibly to get this disease outbreak under control.

If you are not certain of your initial diagnosis of AFB get a second opinion from an experienced registered beekeeper. You can purchase AFB testing kits from beekeeping supply stores.

Different countries have different rules on how to deal with AFB. Contact your local authority to ensure you follow the legal procedures for your area.

Action to be taken immediately if American foulbrood is detected in your hive:

1. Under New Zealand regulations, if you used a hive tool to open your hive, sterilize this in your smoker with a burning flame. Burn any matchsticks if they were used to test for ropiness. Scrub your gloves, smoker and bee suit with a 10 percent bleach solution and then hot wash in soapy water.
2. In the evening return to the hive once the foraging bees have returned and close up all the entrances. Tape up the entrance and any gaps with tape. Pour a cup of petrol over the frames and into the hive and cover with the lid to contain the fumes. The fumes will kill the bees within 15 minutes. Carry the enclosed hive to the burn site, ensuring that no dead bees fall from the hive.
3. To dispose of the infected hive you need to burn all the equipment in a hole or other deep receptacle that will hold all the bees, honey and wax. You do not want any molten wax or honey to be able to drain away, out of the site and for other bees to try and rob this material.
4. Light the fire carefully. With the petrol and wax and dry wood, the fire will be fierce and hot. Ensure that all bees and hive material are burnt. This includes the hive floor, plastic frames and even the hive lid to be extra cautious. Plastic hive parts will create a toxic black flame. Keep a hose or buckets of water nearby for safety.
5. Add other dry timber to the fire to ensure that all hive parts are completely burnt and are turned to a fine ash. This can take several hours.
6. All the ash from the fire should be buried in a hole. Sprinkle a solution of 10 percent bleach solution around and in the site using a watering can to make extra sure that all infective spores are destroyed.

7. If you are a New Zealand beekeeper, notify the Management Agency within seven days upon identifying an infected hive and the course of action taken. You can do this online at www.afb.org.nz. Infected hives should be destroyed within seven days of detection unless other options are permitted under your personal AFB management plan.

I live in the middle of suburbia in a small townhouse. I can't dig a hole to burn my hive what do I do?

If your hive has AFB and you need to destroy it, living in the middle of suburbia can be problematic. The first thing to remember is you still are under a legal obligation to kill the infected bees and destroy your hive. Do not dump your infected bee gear at the dump or transfer station. This is illegal and can also allow bees to rob out the infected honey.

Your legal obligation is to kill the infected hive bees, destroy the bees and all parts of the hive by burning, prevent any runoff of infected wax or honey which could cause robbing, and bury the ashes of the fire under a thick layer of soil. Perhaps you can think outside the square and reach the same outcome?

One suggestion would be to burn the hive at night using a large steel drum to contain all the contents of the hive and to prevent any molten wax or honey from draining away. Bury the cooled ashes under a layer of soil in a large pot or garden area. This method is superior to (and far more legal than) dismantling the hive body and frames and putting them in your wheelie rubbish bin or just storing the hive in your carport until you get around to doing something about it months later.

Do you have a friend who lives in a more appropriate area where you can dig a hole?

Are you a member of a bee club that may be able to help?

If you need to transfer the infected hive to another location to be destroyed, ensure the hive is securely taped so it does not come apart or allow dead bees to fall from the hive.

I have only had my new Flow hive for 18 months. My bee mentor has just told me my hive is infected with American foulbrood. Do I really have to burn it? It cost me over $800 !

I am really sorry but regardless of hive design, or cost, all parts of your hive will need to be destroyed by burning. Especially the plastic frames. Cut your losses and start again. You will be doing your future bees and all hives around your area a favor. As part of a beekeeping community, you have legal and ethical obligations.

How to reduce the chance of spreading American foulbrood into your hive:
- Inspect each and every frame of brood before removing or transferring to another hive
- Always buy new hive ware, avoid second-hand equipment like the plague
- Treat swarms and feral colonies as potential sources of disease
- Do not feed bees honey unless it is from your own hive and you have had no cases of American foulbrood
- Do not feed your bees pollen from another source
- Position your hives to minimize drift
- Attend a workshop on how to recognize American foulbrood and become a certified beekeeper
- Do not interchange frames from different colonies throughout your hives unless they have been thoroughly checked for diseases
- Ensure that all your comb regardless of hive design remains straight and can be individually removed and inspected, both sides, for disease
- Register yourself as a beekeeper so you can be informed of disease outbreaks within your location
- Go through each and every frame of brood at least four times a year to inspect for diseases
- Do not allow hives to be robbed out by others
- Have your hive inspected once a year by an approved beekeeper who has passed a competency test or attend a course and obtain this certification.

Sacbrood

Sacbrood is a virus which is not lethal to the hive but the symptoms look very similar to American foulbrood so it is important to be able to recognize the differences. Young larvae are infected and spring is the time of the year when this disease is most prevalent. In a serious infection you will see a patchy brood pattern and infected larvae lying slumped along the bottom of the cell. In AFB the color change is uniform across the larvae where Sacbrood infected larvae takes on a mottled color of gray, black and brown. The head darkens first. If you drag the infected larvae out of a cell it looks like a sock filled with water. If you puncture the skin, watery fluid runs out. Requeening will help serious cases.

A clear sign that you have American foulbrood in your hive is when the dead larvae ropes out on a matchstick.

The smell emitted from an infected AFB hive is something you will always recognize.

This "shotgun" pattern of the brood frame with many empty or uncapped cells is a common sight with American foulbrood. Also notice the darker sunken cells.

Chalkbrood

Chalkbrood is caused by a fungus and kills the larvae and pupa. When infected the larvae or pupa swell and turn into small cylinders of chalk-like material. Often the first indication of an infection is observing these many colored pieces of chalk outside the entrance of the hive or on the hive floor if you have a screen bottom floor. Housekeeping bees will remove the dead larvae from their cells and discard them outside the hive.

Chalkbrood is more common during periods of cool and damp weather. By increasing the ventilation through the hive (by opening up the

Chalkbrood mummies can be seen on this sticky board which was in position under a hive floor. The chalkbrood are the cream, white and gray-colored cylindrical debris. Hive bees have pulled these infected pupa out of cells and have been removing them from the hive.

bottom board if your hive has a screened bottom, or propping the hive lid open slightly) you can increase airflow through the hive. *If* your hive is situated under a tree or in a gully, moving the hive to a warmer dryer position will help alleviate this disease.

Chalkbrood will not kill a hive but a bad infestation can weaken a hive, leading to robbing or starvation.

Nosema

There are two types of nosema diseases which affect honey bees. Both are protozoan-caused and are spread by spores. These spores attack adult bees. Transmission of this disease is often most prevalent during winter when a colony has been forced inside its hive for long lengths of time. As nosema is a disease of the digestive tract, when the bees are confined and unable to leave for cleansing flights the spores accumulate in the bees' rectum.

Nosema increases the mortality of adult bees. Infections will reduce honey yields and result in a poor population build up. If the queen becomes infected her ovaries are affected and her egg laying capacity greatly diminishes.

If your hive has nosema you may see adult bees crawling around on the ground looking as if they are disorientated, a colony not eating supplementary feeds, a very slow spring build up and fecal streaks on the front of the hive. This can look like someone has splashed the front of the hive with brown paint.

To treat and prevent further infections, avoid transferring comb from infected hives to others, position hives to avoid any drifting of flying bees, prevent robbing of weakened hives and requeen weak hives. A treatment I have been told about but have never tried personally is to add 1 percent acetic acid in sugar syrup on a good flying day inside the affected hive (a tip from Frank Lindsay).

Conclusion

As backyard beekeepers we are lucky to be in a unique position where we can enjoy our hobby without needing to make a living off the bees. Honey does not equal money to us. Having a prosperous and healthy natural hive full of vibrant bees is far more important. Backyard beekeepers have the potential to create a flotilla of beehives following management practices which are focused on the bees' needs rather than honey production or pollination duties. Our job is to listen to the bees, as they know what they are doing.

Walking to the hives. PHOTO: *Alphapix*

AUTHOR TIPS

Here is a suggestion for how a bee-focused backyard beekeeper can manage their hives and care for their bees:

Limit the number of hives in each apiary to five
Why? Bees in the wild prefer to create a hive at least 1.85 miles (3 km) away from another colony to prevent competition for nectar sources. Commercial practices of locating up to 50 hives in one area place stress on the bees and can spread disease. Beekeepers keep a large number of hives in one area as this cuts down on travel, time and apiary registration fees.

Don't use plastic
Would you like to live in a house made from plastic? Many commercial operations use plastic frames and foundation (embossed sheets inside the frames) in their hives, forcing the bees to live, raise their babies and store their food in plastic. Plastic frames have less maintenance and cost for the beekeeper but the bees don't like it. If you have a Flow hive, allow your bees to build natural comb in the all-important brood box to offset the high use of plastic in the honey box.

Don't use foundation. Let your bees build their own unique and magical comb to their own required dimensions
It is the bees' right to build their own wax comb from scratch. After all they have been doing this for hundreds of thousands of years, before man started farming them. Foundation are wax sheets inserted into frames to give the bees a head start on honey production. The foundation is made to a set size for the beekeeper rather than what the bees require. This foundation is reused each year. Over time it becomes black and can harbor disease. Wax also absorbs any synthetic chemicals used in the hive, creating a toxic home for the bees. Would you like the honey you are eating sitting in that?

Her Majesty the Queen is free to roam throughout her hive
Many beekeepers cage the queen in the lower portion of the hive using a wire grill. Follow a more natural method of beekeeping, where all bees in the

colony are free to travel freely throughout the hive. Only harvest combs of honey that have no brood (eggs and larvae) on them.

If you chose to treat for varroa, use only naturally-occurring acids or essential oils.
Synthetic chemicals build up in the wax and now there is an increasing issue of resistance by the varroa mite to many of these chemicals used. Organically managed hives require more time and more visits to treat but it is the only sustainable way forward as we all strive for bees that can manage varroa naturally.

Give your hive a permanent home in a garden
Many commercial hives are moved regularly for pollination duties or so the beekeeper can profit from manuka honey or other single-sourced honey types. This can cause undue stress to the colony and result in large bee losses. Many of the crops that bees are forced to pollinate do not give them adequate nutrition and when doing this task the colony can lose condition rapidly resulting in disease. Urban bees do much better as they have a varied and almost year-round source of nectar and pollen. In urban areas there are also no agricultural sprays or mono crops for our bees to be subjected to.

Encourage drones (the male bees) in your hives, as they are an integral part of any natural colony. They help to pass on important genetics to the new queens
Commercial beekeepers will prevent large numbers of drones in their managed hives as they do not collect any nectar and so are seen as useless.

Let the hive decide when it is the right time to replace their queen
Many commercial beekeepers will routinely kill the queen and replace each year, or second year, to keep honey production high. Other than in cases of emergency, let the hive raise its own queen. This queen will be perfectly adapted to the local environment.

Always leave honey for the bees to feed on during winter when little is flowering.
In a commercial operation, honey is money and white sugar is a cheaper alternative to feed hungry bees.

BIBLIOGRAPHY

Beekeeping: natural, simple and ecological, translated by David Heaf from *Bienenzucht. Naturgerecht einfach und erfolgsicher* by Johann Thür, Imker (Wien, Gerasdorf, Kapellerfeld, 2nd ed., 1946)

Honeybee Biology and Beekeeping, (2013), Dewey M Caron and Lawrence John Connor, Wicwas Press, USA

Honeybee Democracy (2010). Thomas D Seeley, Princeton University Press.

Natural Beekeeping: Organic approaches to modern apiculture, 2nd Edition, (2013), Ross Conrad, Chelsea Green Publishing.

Practical Beekeeping In New Zealand (2011), Andrew Matheson and Murry Reid, Exisle Publishing Ltd, Auckland.

Top-Bar Beekeeping: Organic practices for honeybee health (2012), Les Cowder and Heather Harrell, Chelsea Green Publishing, VT, USA.

Control of Varroa: A guide for New Zealand beekeepers (2007), Mark Goodwin and Michelle Taylor, New Zealand Ministry of Agriculture and Forestry, Wellington, New Zealand.

Randy Oliver, www.scientificbeekeeping.com

Report on the 2015 New Zealand Colony Loss and Survival Survey, (2016) Pike Brown and Linda Newstrom-Lloyd, Landcare Research NZ Ltd, available at www.landcare-research.co.nz/science/portfolios/enhancing-policy-effectiveness/bee-health

beeinformed.org/results/the-bee-informed-partnership-national-management-survey-2014-2015

newsandfeatures.uncg.edu/kaira-wagoner-honey-bee-research

Jandricic, Sarah and Otis, Gard, 'The Potential for Using Male Selection in Breeding Honey Bees Resistant To Varroa Destructor', *Bee World*, 84 (4): 155–164, 2003

mysare.sare.org/mySARE/ProjectReport.aspx?do=viewRept&pn=FS12-261&y=2013&t=1

First published in 2017 by New Holland Publishers Pty Ltd
London • Sydney • Auckland

The Chandlery Unit 704 50 Westminster Bridge Road, London SE1 7QY, United Kingdom
1/66 Gibbes Street, Chatswood NSW 2067 Australia
5/39 Woodside Ave Northcote, Auckland 0627 New Zealand

www.newhollandpublishers.com

Copyright © 2017 New Holland Publishers Pty Ltd
Copyright © 2017 in text: Janet Luke
Copyright © 2017 in images: Janet Luke except where noted

All rights reserved. No part of this publication may be reproduced, stored in a retrieval system or transmitted, in any form or by any means, electronic, mechanical, photocopying, recording or otherwise, without the prior written permission of the publishers and copyright holders.

A record of this book is held at the British Library and the National Library of Australia.

ISBN 9781869664565

Group Managing Director: Fiona Schultz
Publisher: Christine Thomson
Copy Editor: Gordana Trifunovic
Project Editor: Liz Hardy
Designer: Andrew Davies
Production Director: James Mills-Hicks
Printer: Times International Printers, Malaysia

10 9 8 7 6 5 4 3 2 1

Keep up with New Holland Publishers on Facebook
www.facebook.com/NewHollandPublishers